T0201525

Bruno Latour

Key Contemporary Thinkers

Lee Braver, *Heidegger*
John Burgess, *Kripke*
Claire Colebrook and Jason Maxwell, *Agamben*
Jean-Pierre Couture, *Sloterdijk*
Rosemary Cowan, *Cornel West*
George Crowder, *Isaiah Berlin*
Gareth Dale, *Karl Polanyi*
Colin Davis, *Levinas*
Oliver Davis, *Jacques Rancière*
Gerard de Vries, *Bruno Latour*
Reidar Andreas Due, *Deleuze*
Edward Fullbrook and Kate Fullbrook, *Simone de Beauvoir*
Andrew Gamble, *Hayek*
Neil Gascoigne, *Richard Rorty*
Nigel Gibson, *Fanon*
Graeme Gilloch, *Siegfried Kracauer*
Graeme Gilloch, *Walter Benjamin*
Phillip Hansen, *Hannah Arendt*
Sean Homer, *Fredric Jameson*
Christina Howells, *Derrida*
Simon Jarvis, *Adorno*
Rachel Jones, *Irigaray*
Sarah Kay, *Žižek*
S. K. Keltner, *Kristeva*
Valerie Kennedy, *Edward Said*
Chandran Kukathas and Philip Pettit, *Rawls*
Moya Lloyd, *Judith Butler*
James McGilvray, *Chomsky*, 2nd Edition
Lois McNay, *Foucault*
Dermot Moran, *Edmund Husserl*
Michael Moriarty, *Roland Barthes*
Marie-Eve Morin, *Jean-Luc Nancy*
Stephen Morton, *Gayatri Spivak*
Timothy Murphy, *Antonio Negri*
William Outhwaite, *Habermas*, 2nd Edition
Kari Palonen, *Quentin Skinner*
Ed Pluth, *Badiou*
John Preston, *Feyerabend*
Chris Rojek, *Stuart Hall*
Severin Schroeder, *Wittgenstein*
Anthony Paul Smith, *Laruelle*
Dennis Smith, *Zygmunt Bauman*
Felix Stalder, *Manuel Castells*
Georgia Warnke, *Gadamer*
Jonathan Wolff, *Robert Nozick*
Christopher Zurn, *Axel Honneth*

Bruno Latour

Gerard de Vries

polity

First published in 2016 by Polity Press

Polity Press
65 Bridge Street
Cambridge CB2 1UR, UK

Polity Press
350 Main Street
Malden, MA 02148, USA

ISBN-13: 978-0-7456-5062-3 (hardback)
ISBN-13: 978-0-7456-5063-0 (paperback)

A catalogue record for this book is available from the British Library.

Library of Congress Cataloging-in-Publication Data

Names: Vries, Gerard de, 1948-
Title: Bruno Latour / Gerard de Vries.
Description: Cambridge, UK ; Malden, MA : Polity Press, 2016. |
    Includes bibliographical references and index.
Identifiers: LCCN 2016001169 | ISBN 9780745650623 (hardback : alk.
    paper) | ISBN 9780745650630 (pbk. : alk. paper)
Subjects: LCSH: Latour, Bruno. | Science–Social aspects. | Science and
    civilization. | Science–Philosophy.
Classification: LCC Q175.46 .V75 2016 | DDC 303.48/3–dc23 LC record
    available at http://lccn.loc.gov/201600116

Typeset in 10.5 on 12 pt Palatino
by Toppan Best-set Premedia Limited
Printed and bound in the UK by CPI Group (UK) Ltd, Croydon, CRO 4YY

For further information on Polity, visit our website: politybooks.com

Contents

Preface

"When men cannot observe, they don't have ideas; they have obsessions," V. S. Naipaul wrote. The modern philosophical tradition holds observation in high regard; nevertheless it is obsessed by a worldview that feeds off dualities – between humans and nonhumans, nature and society, facts and values, science and politics. Bruno Latour wants us to observe better, with a finer resolution. To become more attentive, to redescribe the world we live in, and to better understand our current predicament, he introduced ethnography and comparative anthropology as vital methods for philosophy. Latour is an 'empirical philosopher'.

This introduction to his work follows Latour in his footsteps, both as an ethnographer and as a philosopher. The light tone of much of Latour's writing may easily conceal its profundity. Latour takes issue with much of what we take for granted as intuitively evident. By following Latour's moves closely and by providing some background from science studies, philosophy and sociology, to show to what extent and in what sense Latour's work stands out against the tradition, I hope to ease access to what Latour claims to be a richer vocabulary to account for who we are and what we value, that is, a better, fairer common sense.

Latour is a prolific writer on an amazingly varied set of topics, so some selection was inevitable. To introduce his empirical work, Latour's studies on science, law and religion will be discussed in detail; they led to substantial philosophical innovations. Latour's philosophy – his thoughts on science, on actor-network theory, cosmopolitics and his anthropology of the Moderns – is introduced roughly in the order in which they took shape. But this is not an intellectual biography; the historical and intellectual context in which Latour's thoughts evolved is only touched upon. That also holds for the reception of his work. This is an introduction to Latour's philosophy; not to science studies as a discipline, nor to the work of those who have followed Latour, used his ideas, or thought they did.

To write about a living author is an unquiet affair. With the advantage of hindsight it becomes apparent that Latour's work has been driven by a coherent heuristic. But those who followed his work were often puzzled when he took his thoughts to new levels and new domains or when he introduced conceptual innovations. We met in the early 1980s and stayed in contact ever since. Time and again he forced me to rethink his position, as well as my own.

I want to thank Bruno Latour and my Dutch friends and colleagues Huub Dijstelbloem, Rob Hagendijk, Hans Harbers, Josta de Hoog, Noortje Marres and Annemiek Nelis for their comments on the draft of this book. I'm also very grateful to John Naughton for his comments and for helping me out with the subtleties of the English grammar. As always, my gratitude to Pauline extends far beyond her comments on my writing.

Abbreviations

For full bibliographical details see the References.

AIME   *An Inquiry into Modes of Existence – An Anthropology of the Moderns*
AR   *Aramis or the Love of Technology*
CB   *La Clef de Berlin et autre leçons d'un amateur de sciences*
CM   *Petite réflexion sur le culte moderne des dieux faitiches*
FG   *Face à Gaïa*
ICON   *Iconoclash – Beyond the Image Wars in Science, Religion, and Art*
IRR   *Irreductions* (part 2 of *The Pasteurization of France*, cited by paragraph number)
LL   *Laboratory Life – The Social Construction of Scientific Facts*
LL2   *Laboratory Life – The Construction of Scientific Facts* (2nd edition)
ML   *The Making of Law*
MTP   *Making Things Public – Atmospheres of Democracy*
NBM   *We Have Never Been Modern*
PF   *The Pasteurization of France*
PH   *Pandora's Hope – Essays on the Reality of Science Studies*
PN   *Politics of Nature – How to Bring the Sciences into Democracy*
PVI   *Paris ville invisible*
RAS   *Reassembling the Social – An Introduction to Actor-Network-Theory*
REJ   *Rejoicing – Or the Torments of Religious Speech*
SA   *Science in Action*
SPI   *The Science of Passionate Interests*

1

Empirical Philosophy

A truly bewildering *Wunderkammer*. The collection of Bruno Latour's publications brings an early-modern cabinet of curiosities to mind. Their subject matters range from laboratory life in Nobel Prize winner Roger Guillemin's Salk Institute (LL); the shared history of microbes, microbiologists and society (PF); the tragic fate of an innovative public transport system (AR); file handling and the passage of law at the French supreme court for adminis-trative law, the *Conseil d'État* (ML); geopolitics in the epoch of the Anthropocene (FG); religion (REJ); economics (SPI); ethnopsychia-try (CM); modernity (NBM, AIME); to Paris (PVI), politics (PN, MTP) and philosophy of science (SA, PH). Pick up any of Latour's books and you will be guided again through a maze of surprising connections – from the technical details of a rotary motor to meet-ings at the Transportation Ministry and a photo-op with French Prime Minister Jacques Chirac; from Pasteur's laboratory in Paris to a farm in Pouilly-le-Fort; from the *Conseil d'État* in session to its mail room and to the file folders, stamps and paperclips in a sec-retary's cubicle; from a dialogue between lovers to Fra Angelico's fresco of the empty tomb in Florence; or from the Salk Institute's lab benches to the hectic travel schedule of its boss on his way to meet an endless array of colleagues, firms and high-level civil servants.

What's the point? Is there any order in this confusing, rambling, seemingly boundless list of subjects, actors, institutions and places? Who's interested in a High Court secretary's office paraphernalia or in the smile on the face of a politician sitting in a prototype of a public transport system? Why doesn't Latour stick to science and technology, the study of which brought him international fame; why did he fan out to other topics? And why doesn't he take the trouble to sort the various aspects of what he's talking about into neat categories, to leave technical details to scientists and engineers, legal matters to jurists, so that social scientists can focus on organizational matters, institutional relations and politics, after which philosophers can sit down to discuss foundational and methodological issues, the crumbs that fall from the other disciplines' dinner tables?

To find our way in the world, to understand the modern world we live in, Latour claims we have to abandon the intuitions and explanatory ideals we have been trained to hold dear. The world does not present itself in pre-packed items that nicely fit into the pigeonholes of the established scientific disciplines. An education in law may have prepared lawyers for carefully reading texts and discussing legal subtleties, but if secretaries and court-clerks run out of file folders and paperclips, court documents will get messed up, chaos will emerge and before long the process of administering justice will have come to a full stop. So if we are interested in what lawyers are doing and how justice is administered, we better start taking an interest also in such seemingly trivial material aspects of legal practice as file folders and other office paraphernalia. Is it really possible to understand anything about societies without taking technology into account? Amazingly, the topic is not covered by sociology. Pick up any sociology textbook and you will see that sociologists are trained to study human groups, institutions, cultures and maybe the impact of technology on society, but not how technology makes up a substantial part of the fabric of society. So, Latour boldly claims, to become realistic about society, sociology has to be reformed. Can one truly understand modern science if one neglects the fact that a winner of the Nobel Prize in Physiology or Medicine will have spent long days not only at his lab, in hospitals and on academic conferences, but also in meeting rooms to discuss his work with patent lawyers, representatives of pharmaceutical firms and government officials? No. So we better start rewriting the usual stories about science. Why should one respect the established boundaries of scientific disciplines if scientists themselves keep trespassing them time and again? Engineers mix with politicians for work, not for pleasure; chemists, biologists and climate scientists discuss ecological problems with government representatives who probably have been educated as economists,

lawyers, or as policy analysts. Only social theorists and philosophers tend to rigidly guard the borders of their fields.

As a consequence of his lack of respect for disciplinary boundaries, Latour's work is difficult to label. No wonder bookshops find
it difficult to decide where to place his books on their shelves. In
Paris, you will find most of them in the Social Sciences section; in
Oxford and Cambridge they are stored under History and Philosophy of Science; in Amsterdam in the Philosophy corner. His papers
are published in a wide variety of journals. To locate some of his
other work, you may even have to travel to a museum – to the
Karlsruhe *Zentrum für Kunst und Medientechnologie* (ZKM) where
Latour has curated exhibitions, or to the *Centre Pompidou* in Paris
where he organized a series of conferences – or to visit his website
www.bruno-latour.fr to find a 'sociological web opera' posted next
to a video of a re-staged debate between Durkheim and Tarde, with
Latour impersonating the latter. The fruits of Latour's labours are
not easy to categorize.

But first impressions are deceptive. What may appear as a hotch-
potch of projects that lacks disciplinary rigour is driven by a clear
intent, namely to describe science, law, politics, religion and other
key institutions of the modern world in a new way. Latour claims
that several of the established conceptual distinctions used to
demarcate modern institutions – e.g. nature versus society, and
facts versus values – provide at best little guidance to understanding what goes on in science, law, politics and religion and more
likely will lead us astray. To articulate the nature of the world we
live in, to get a more realistic view, we need to *redescribe* these
institutions, their values and the ways in which they differ from
one another. In spite of the fact that most of Latour's academic
papers have been published in social science journals, this intent is
sufficient reason to conceive Latour primarily as a philosopher –
although one of a distinctive kind.

Traditionally, philosophers have conceived their task as finding
a point on solid ground that allows a perspective on the world as
it is, that is, to see reality, essences, behind confusing appearances.
Plato's allegory of the cave nicely captures the ambition. In contrast
to the prisoners who have been chained in a cave for all their lives,
who are able only to look forward and who take the shadows cast
on the wall in front of them for reality, the philosopher is like the
one who is freed from his fetters, who raises and turning around
is confronted with the things outside the cave that cast the shadows
on the wall, and who comes to understand that what he had seen

before was all a cheat and an illusion, and that now he has turned
to more real things he can see more truly (Plato *Republic*: 7.514).

Over time, philosophers have adopted a more modest attitude.
The rise of the sciences has forced them to reconsider their role. As
Foucault observed:

> [f]or a long time one has known that the role of philosophy is not to
> discover what is hidden, but to make visible precisely what is visible,
> that is to say, to make evident what is so close, so immediate, so
> intimately linked to us, that because of that we do not perceive
> it. Whereas the role of science is to reveal what we do not see, the
> role of philosophy is to let us see what we see. (Foucault 1994:
> 3.540–541)

The sciences aim to inform us about what is hidden from view –
e.g. what goes on in a distant star system, or in the brain of an
Alzheimer patient – and to explain what we see in terms of underlying structures and processes. In contrast, philosophy tries to
provide *redescriptions* of what is close to us: the world we live in
and relate to, our social and moral intuitions, and our notions of
who we are. So we may find Foucault (1979) opening our eyes for
a much wider range of ways in which the conduct of a person is
controlled in modern society, namely by pointing to new forms of
discipline and punishment, that is, forms of power that have been
around ever since about the early nineteenth century but that went
unnoticed because we used to understand power only to refer to
"every chance to carry through one's own will, even against resistance" (Weber 1972b [1922]: 28).

To single out redescription as the specific role for philosophy is
certainly not an aberrant preference of some French philosophers
alone. For example, Wittgenstein (1969 [1952]) also declared that
"[w]e must do away with all *explanation*, and description alone
must take its place" (PU §109). "We want to *understand* something
that is already in plain view" (PU §89). The technique he suggested
differs from Foucault's approach. For Wittgenstein, careful description of language is the preferred way to provide *Übersichtlichkeit*, a
'perspicuous survey' that helps to untie the knots in our understanding and to resolve philosophical perplexity. A perspicuous
representation will "bring about the understanding which consists
precisely in the fact that we 'see the connections'." "Hence," he
added, "the importance of finding *connecting links*" (Wittgenstein
1993: 132).

The similarity with Latour's intent is as remarkable as the difference. While at one point Wittgenstein (1998: 45e) wondered "if we use the ethnological approach, does that mean we are saying philosophy is ethnology?" – to further limit his attention to describing language games – Latour rose from his armchair to grab the bull by its horns. To do philosophy, to actually trace the connecting links and to learn to see what we see, Latour got engaged in empirical field studies, in ethnography.

1.1   Making Paris visible

Latour's intent and approach to philosophy may become clearer by discussing what at first sight is the most un-philosophical book he has ever published, *Paris ville invisible* (1998), co-authored with photographer Emilie Hermant. Because of its title and design – hundreds of photos of Parisian sites, with text interspersed, printed in a coffee table format – to acquire this book you may have to go to the travel section of a Parisian bookshop, where you may find it next to glossy books about romantic Paris and the *Guide Michelin*. But the rushed tourist who has picked up the book on his way home will likely be disappointed when he unwraps his souvenir. He has bought a treatise on philosophy and social theory. Discovering that the book was later turned into an interactive website will probably add to his chagrin.

In *Paris ville invisible*, the grandiose task of attaining *Übersicht*, of perspicuously surveying the world as a whole, is reduced to the more mundane one of capturing the whole of Paris at a glance. Where do we have to go to accomplish the task of perspicuously representing Paris? From which fetters do we have to free ourselves? Do we have to escape from the Earth, to get a view of the whole of Paris from a satellite? When we look at the image offered by Google Earth, we may indeed see 'the whole of Paris' at a glance. But except for the word 'Paris' being superimposed on the picture on our screen, we might easily have taken it to depict any other city. On the scale that captures Paris as a whole, the trained eye may spot the curves of the Seine, but very little else. So we may decide to take another tack, to go to Paris to join Latour and Hermant on their visit to the *Samaritaine*, the department store near Pont Neuf, which – before it was closed in 2005 for security reasons – proudly advertised itself with the slogan "You can find anything at the *Samaritaine*". On the top floor of the old store was a panorama. One could see a lot of Paris from this spot. Binoculars were available for visitors and there was a huge circular table with engraved arrows pointing to Parisian landmarks drawn in perspective to help orientation. So is this the place where one might see the whole of Paris?

Unfortunately, no. As Hermant's photos show, smog from exhaust fumes veils the view. Moreover, the panorama fails to locate the *Centre Pompidou* and the impressive architecture of *La Défense*, and where the panorama promises tree-covered hills to be visible in the northeast, as Latour and Hermant note, one vaguely sees only endlessly more buildings. Set up in the 1930s, the panorama no longer corresponds to the city that spreads out before us. So, in spite of the available binoculars on the top-floor of the *Samaritaine*, Latour and Hermant suggest that to really see the *Sacré Coeur*, we had better get the metro to Montmartre.

On arrival at metro station *Abbesses*, however, another disappointment waits for us. Once we have left the station, we get lost in the maze of little streets in Montmartre. Where are we? The answer comes from the Michelin map of Paris that we have been carrying around. We look at the nearest road sign – it reads 'Rue la Vieuville' – and soon we have found the same words on the Michelin map. Now we know where we are. On the map, we have the entire eighteenth *arrondisement* at a glance. A minute later we know how to walk to the basilica. However, we also suddenly realize that by further unfolding the map, we can have the whole of Paris at a glance! Plato was wrong. To see the whole of Paris at a glance, we need to divert our attention *away* from the city, away from reality, and to look at the map. To take it all in at once, to see it at a glance, to see its structure, Paris first had to become small. What we set out to see – the whole of Paris – we have been carrying around all day in our pocket.

That is, provided a lot of *work* has already been done. We could find the place where we got lost only because the mapmakers at Michelin meticulously did their job and because the Paris municipality's road-maintenance service has taken the trouble to attach a nameplate to the wall. How did the road-maintenance service personnel know which plate should be attached to this particular wall? Obviously, they too had a map. Another official department, the *Service Parcellaire*, which keeps a detailed record of all cadastral data, has provided it. When we join Latour and Hermant to visit the *Service Parcellaire* at Boulevard Morland, we learn that it gets its information from the *Service Technique de la Documentation*

Foncière, the Ordinance Survey Department, which hires small teams of surveyors to carefully measure the dense fabric of Paris and which gets the official names of streets, once they have been approved by the mayor of Paris, from the *Service de la Nomenclature.*

Our little excursion has taught us a profound philosophical lesson. Paris, reality, cannot be captured at a glance from a single, exclusive point, but the efforts of cartographers, technicians and civil servants, *make* it visible. Their coordinated efforts materialize the conditions that will allow documents, such as a map, to apply to the world, thus helping those concerned to find their way around to know where they are, and to see what they are seeing. If a philosopher sets out "to make visible precisely what is visible", he has to trace the long cascade of activities and techniques that enable us to see what we see. Step by step they have transformed the *terra incognita* we have been plunged in – what in *Paris ville invisible* Latour calls the 'plasma' – into an ordered reality, while simultaneously delivering the means to represent, to map, its order. What was left for the mapmakers at Michelin to do was carefully to translate the official maps that resulted from the coordinated work of several municipal services into a format suitable for tourists to read and to carry around. Without these services, the Michelin mapmakers too would get lost.

Once we have digested this lesson, we will soon find out that there exist numerous ways of making Paris visible. There are multiple Parises within Paris. So let us join Latour and Hermant for another tour. At *SAGEP* the water supply is controlled. Computers survey the intake and consumption of fresh water. A lot of data need to be processed. But in spite of what the Latin word 'data' suggests, these data are not given, but need to be obtained. Sensors duly record the water flows, but to indicate where leakages may have occurred their readings need to be processed and compared to normal values. Because water will spend on average six hours in the system before being consumed, it is also necessary to anticipate demand. Statistics and weather reports help to predict upcoming consumption. *SAGEP* also knows that at the end of an European Cup football match, thousands of toilets will be flushed almost simultaneously. To control the water supply of a big city not only requires processing data from a lot of pipes, pumps and sluices, but also sociological insight into the habits of the population.

Similarly, traffic controllers, market researchers, statistical bureaux, all rely on techniques to obtain data and to translate them into images on a scale that fits the screen of a computer or a sheet of paper. All of them make Parisian lives possible and visible at the same time. Where would you find 'the market' if it wasn't made possible to compare the prices of commodities? Somebody has to collect and process them for you, to make visible what 'the market' is doing. Even local supermarkets contribute to this task by carefully arranging products on their shelves and by making the price and the content of products visible so that clients can make comparisons before loading their shopping carts. There is no market without devices that make what's on the market visible.

Okay, one may think, this may perhaps apply to social reality and technology, but surely not to the natural world? It takes little effort to realize that technologies like the water supply system are constructed. And because social facts require human agreement and institutions for their existence, the same goes for acknowledging that the building blocks of social reality are constructed. But surely, natural facts are 'brute facts' that require no human institution for their existence. Of course, to *state* a brute fact requires the institution of language, but the fact stated needs to be distinguished from the statement of it (Searle 1995: 2). One should not confuse the map with the territory. In contrast to social facts, brute, natural facts – say the fact that the sun is 150 million kilometres from the Earth – existed long before any human being appeared on Earth. It would be strange to think that in this case human construction work would be involved.

Alas, it's more complicated. Join Latour and Hermant for yet another trip, this time to the biology department of the *École de Physique et Chimie de Paris*. Here, Dr Audinat has managed to make visible the activity of a single neuron in a rat's brain. 'Molecular and Physiological Diversity of Cortical Nonpyramidal Cells', the paper in the *Journal of Neuroscience* in which he and seven colleagues have published their results (Cauli et al. 1997), carefully documents all the steps that were necessary to do the job. On one of its pages, a photo showing a neuron – a 'layer V FS cell' – is juxtaposed with a graph of its electric potential and a photograph of an electrophoresis gel that bears the marks of molecules synthesized by this neuron. We see the anatomy, the electric potential and the molecular biochemistry of a neuron at a glance – that is, if we read the explanation of the pictures published on the bottom of the page and in the body of the article.

To produce these pictures has required a lot of preparatory work. A rat had to be decapitated, its brain extracted; fine slices of the brain had to be cut with a microtome; these slices had to be framed before being put under an infrared microscope and searching through all of them, a neuron had to be identified. Once this had been achieved, microelectrodes were brought into contact with the cell to obtain the electric potential and to read its electric activity from an oscilloscope. At the end of the recording, the content of the cell was aspirated in a micropipette, transferred to another laboratory, and prepared for introduction into a PCR-machine to obtain the molecular biochemistry of the neurotransmitters. At this point Audinat and his colleagues have obtained their data – which, again, were not 'given', but obviously required a lot of work and skill to be produced. They have finally succeeded in making the activity of a rat's neuron visible. Does their paper refer to the activity of a neuron in a rat's brain? Do the results truly represent events in a rat's brain? Yes, after carefully reading the article the readers of the *Journal of Neuroscience* will be convinced they do – as have been the referees who had to evaluate the paper before the journal's editor decided to publish a second version, revised on the basis of the referees' comments. So, did Audinat and his colleagues state 'brute' facts? To accept that they did, we have to ignore all the steps they had to take to produce the facts they state in their paper – some quite brutal indeed, like decapitating animals. Obviously, to do so they had to rely on much more than language alone. They had to carefully prepare rat-neurons to make their activity visible and representable. To understand what it means 'to refer to' the activity of a neuron, 'to represent' the facts, we have to follow all the links that the group has established in their research – just as we have to understand all the work that the various services had to perform before the Michelin map of Paris became a reliable guide for finding one's way through Paris, and to grasp all the efforts that had to be undertaken before *SAGEP* knows the current status of the water supply in the complicated system of pipes and sluices it oversees.

In *Paris ville invisible* we encounter Latour in action. Visiting an amazingly varied collection of sites, he interviews the people he encounters on his way, always curious to learn the minute details of their work, the way their work is linked to that of others, and the ways they handle, compare, and translate the information they receive in the form of documents, instrument readings, graphs, or in whatever shape it happens to arrive in their offices, into a form suitable to their own purposes. He carefully listens to their accounts of what is going on, the technical and organizational problems they have to cope with, and the alternatives that may have existed but which they did choose not to follow. He is engaged in ethnographic fieldwork, in empirical, anthropological research – not of some remote tribe on some Pacific island, but of the people working at *SAGEP*, a biochemistry laboratory in Paris or in California, the French High Court for administrative law, or at any other site that might have attracted his curiosity.

While scribbling observations in his unreadable handwriting in little notebooks, to be transposed later to other notebooks, to finally find their way into the case studies that populate his work, Latour simultaneously answers some of the key questions of philosophy. *From which point can we see the true structure of reality?* Wrong, badly framed question. There is no such point, but nevertheless we may study how reality is *made* visible. *How do you know that the pictures that are eventually produced correspond to reality?* Again, wrong, badly framed question. The pictures that are produced do not correspond to reality, but to other pictures. If for whatever reason a problem arises, Michelin will first check its maps against the ones produced by the *Service Parcellaire*. If the Michelin mapmakers find no fault, they will inform this service, which again will first check its own records. If the *Service Parcellaire* finds its records to be in order, it will inform the *Service Technique de la Documentation Foncière*, which ultimately may decide to send out a fresh team of geodesians to re-measure the site and to draw a new map, which will be processed all the way up to the mapmakers at Michelin, who will dutifully issue a new edition of their map. *But aren't we in this way confusing the map with the territory?* Again wrong question. It suggests that we are dealing only with an *epistemological* issue, that is an issue about the *knowledge* supposedly contained in a map or a research paper. The real issue lies on the other side of the equation. Epistemologists tend to think much too naively about 'reality'. They conceive it as something given, out-there, as a territory waiting to be discovered and to be mapped. However, as we have just seen, it takes a whole array of preparatory actions to make reality visible, measurable and to interact with it. To observe the activity of neurons in rats' brains, Dr Audinat didn't passively watch an animal, but decapitated a rat to meticulously prepare slices of its brain to make a neuron's activity visible. What we're primarily dealing with when trying to understand what scientists – or mapmakers for that matter – are involved in, are first of all *ontological* questions, that is to say questions about what, and in what way, something has to *be*, before it can properly be called 'objective', 'visible reality'.

Before a scientific paper, a document, a map can refer to something out-there, to reality, reality has to be transformed from a *terra incognita*, from a 'plasma', to a territory that is *made* visible. To 'passively' observe facts one has first to 'actively' invest in reality (Fleck 1979 [1935]). *But philosophers of science surely must have been aware that setting up a scientific enquiry involves a lot of work, haven't they?* Indeed. Of course, they knew this. But once they started to writing up their accounts of what makes scientific knowledge great, they seem to have forgotten this. *How come?* – We'll get to that later.

This is what Latour is up to: while filling up his little notebooks, he is replacing epistemological questions that have dominated most of the philosophical tradition by ontological ones. And in contrast to most philosophers of the past who engaged their inquiries in an armchair, to answer his philosophical questions Latour goes out to do ethnographical research. For want of a better name, we may call him an 'empirical philosopher'. Neglect his empirical work and you will completely lose his philosophy; disregard the philosophical intent and you will be bogged down in a bewildering set of disparate books and papers.

1.2 The path towards 'empirical philosophy'

It took quite some time before Latour decided to come out as a philosopher.

Born in 1947 in Beaune (France), in a family that has owned *Maison Louis Latour* since 1789, world-famous growers and merchants of the finest Burgundy wines, Bruno Latour left the future of the wine-business to his brother to attend Jesuit school and the University of Dijon for a master's degree. In 1972 he gained a First in the *agrégation de philosophie* (a national exam). In Ivory Coast, while engaging in 'cooperation', a sort of French Peace Corps, an alternative available at the time for military service, he completed his *thèse de troisième siècle* (PhD) *Exégèse et Ontologie* (1975, l'Université de Tours). It includes a close reading of *Clio, dialogue de l'histoire et de l'âme païenne*, an abstruse work of the poet, essayist and philosopher Charles Péguy that provided the material for '*Les raisons profondes du style répétitif de Péguy*', a paper he gave to the Péguy centennial conference in 1973. It was published in 1977 as '*Pourquoi Péguy se répète-t-il? Péguy est-il illisible?*' – 'Why does Péguy repeat himself? Is Péguy unreadable?' – his first academic paper. In 1987 he obtained the *habilitation à la direction des recherches* at the *École des Hautes Études en Sciences Sociales*, a second PhD.

Ivory Coast got Latour on the road to ethnography. Invited to contribute to a study on the problems encountered when replacing white executives with local, black Ivory Coast managers, Latour decided to study how 'competence' was conceived in the industrial milieus of Abidjan. Based on extensive, two to three hours interviews with about 130 persons, assisted by Amina Shabou, Latour identified the discourses about competence of four distinct groups: European executives and lower middle-class whites; black Ivory Coast executives; directors of big multinational corporations; and black workers. Speaking in terms of 'race', the 'nature' and 'culture' of the native population, or in terms of features like 'courage', 'talent', and 'bad faith', the first two groups were found to offer accounts of the 'African mind' to explain why the black population was still not up to managing modern industrial enterprises. These accounts, however, suffered from one problem, Latour observed. They provided explanations for a fact that was far from evident. Although there was an abundance of anecdotes (conveyed in the same discourse), statistical data about native incompetence were non-existent. And when interviewing directors of multinational firms, Latour found this group to conceive competence primarily as a matter of proper recruitment, training and management. Instead of referring to the 'African mind' of the local population, the higher levels of the Abidjan business community talked about 'organization charts', 'proper stimulation' and 'promotions'. So, if there was incompetence among Ivory Coast workers at all, with the proper means, perhaps it could be overcome. Latour therefore turned his attention to education. Interviewing white teachers, he noted again complaints about the features of the 'African mind' that prevented African pupils from meeting 'French levels of competence'. For example, he found teachers at the *Lycée Technique* in Abidjan reporting their pupils to be unable to read technical drawings as representing three-dimensional objects, obviously a serious deficiency for future technicians. When he interviewed the pupils and started looking into school practices, however, Latour found a much simpler explanation. The school system (an exact copy of the French system) introduced engineering before students had done any practical work on engines. Since most of the pupils had never seen or handled an engine before, it was not surprising that the interpretation of technical drawings presented them with quite a puzzle. The cause of the problems the students had with reading technical drawings was not their 'African mind', but the lack of appropriate *connections* required to interpret such drawings. Exporting the French school system to Africa without exporting the many links to engines that French pupils have established even before entering school, made boys in Abidjan 'incompetent'.

Latour had made a small but significant move. What had started as a rather dull sociological enquiry, and halfway through might have turned into a 1970s style critique of ideology and critical social science, suddenly had become quite another kind of enterprise. Competence is not a state of mind that precedes successful action, Latour concluded in *Les idéologies de la compétence en milieu industriel à Abidjan*, the report in which he and Shabou (1974) describe their findings. Competence is not some hidden, given mental entity. One is competent if one controls a system – a machine, an organization, a flow of documents – from beginning to end, that is, if one has the information and the resources ready to adapt and to act capably, that is: if one knows what to do next. Competence should be analysed as set of *links*, as a network of connections that provides a *key* for what to do next, for further action.

In Ivory Coast, Latour's empirical research programme took off. However, the crucial move that led him to conclude that 'competence' should not be conceived in mental and cultural terms but in terms of a set of links, Latour had already anticipated in philosophical terms. In 1972, before leaving France for Ivory Coast, Latour had taught for a short time at a *lycée* in Gray, on the borders of Franche-Comté and Burgundy. As he was to recount in *The Pasteurization of France* (1988), one day at the end of autumn, on his way from Dijon to Gray, he was forced to stop, "brought to my senses after an overdose of reductionism". "I [. . .] simply repeated to myself: 'Nothing can be reduced to anything else, nothing can be deduced from anything else, everything may be allied to everything else' " (PF: 162–3). In Ivory Coast, he applied this 'irreductionist' principle. Instead of reducing the (supposed) lack of competence of blacks workers to their 'African mind', Latour *redescribed* what competence *is* in terms of links, alliances. In Ivory Coast, he started to do philosophy with empirical means.

After having defended his PhD thesis at the University of Tours in 1975, Latour continued on the empirical path. The endocrinologist Roger Guillemin had generously allowed Latour to spend two years (1975–77) at the Salk Institute, San Diego, to study the competences of scientists in a biochemistry laboratory. The lead question of the empirical work Latour would do in California came straight out of his work in Africa: what would happen if the field methods used to study Ivory Coast pupils and workers were applied to first-rate scientists? *Laboratory Life – The Social Construction of Scientific Facts*, the book based on his fieldwork in San Diego, co-authored with the British sociologist Steve Woolgar, was published in 1979.

Latour had met Woolgar in 1976 at the first conference the *Society for the Social Study of Science* organized and had invited him to come to visit the Salk Institute. Woolgar opened up a new resource for Latour, ethnomethodology, the study of the accounts people give of their lives to make sense of their actions and relations and to organize their everyday life. Developed by Garfinkel (1984 [1967]), the approach basically consists in asking people a very simple question – 'what are you doing?' – and – much more difficult – to systematically refrain from offering descriptions and explanations of actions in terms of the schemes taught in social theory classes. Ethnomethodology shifts the attention of social science away from questions about explanations, that is, questions about *why* something happens, to ontological ones, that is, questions about *what* is going on. Latour discovered that he had been practising ethnomethodology for years. This was the approach he had used when interviewing schoolboys and black workers in Abidjan, and this was the question that – not being versed in science – he was raising when meeting the scientists at the Salk laboratory to make sense of what was going on.

Laboratory Life was widely received as a publication in the emerging discipline of 'social studies of science'. Latour was included in the ranks of Barry Barnes, David Bloor, Harry Collins, Karin Knorr-Cetina, John Law, Mike Lynch, Donald MacKenzie, Steven Shapin and a dozen other Young Turks who had decided to turn their backs on established logical-empiricist philosophy of science to take 'a social turn', that is, to study the sciences empirically, by sociological and ethnographical methods. Although there had been sociologists who had studied science as an institution before, it had taken Kuhn's *The Structure of Scientific Revolutions* – originally published in 1962, but more explicitly in its second, enlarged edition (Kuhn 1970) – to provide the basis for the conception of the social character of scientific knowledge. 'Social studies of science' set out to fill in the details by providing an avalanche of sociological studies of scientific controversies and ethnographies of modern laboratories. The way controversies were closed and scientific facts became established was explained not by referring to the available evidence and methodological rules, but in terms of social causes and processes.

Few realized at that time that Latour was on a different trajectory than most of his colleagues in science studies, who self-consciously framed themselves as sociologists, and who were proud not to be philosophers. In fact, Latour doesn't seem to have been aware of the difference himself. It would take several years before he started to deny explicitly that, like the others, he was out to provide socio-logical explanations for scientific knowledge and technological artefacts. Perhaps *Les Microbes*, published with a philosophy-heavy second part, *Irréductions*, in 1984, should have rung some bells, but it was almost completely neglected until *The Pasteurization of France*, its English translation, became available in 1988. To most non-Francophones, the 1986 postscript to the second edition of *Labora-tory Life* first indicated the upcoming differences. In sections with titles such as 'How Radical is Radical?', 'The Place of Philosophy', and 'The Demise of the 'Social', Latour and Woolgar boldly set out why they had omitted the term 'social' from the new edition's subtitle. They pointed out that by explaining the construction of scientific facts in terms of social causes or processes, social study of science – while proclaiming the need to demystify realist epis-temology among natural scientists – had un-reflexively adopted a realist attitude for its own work and had naively misunderstood the nature of ethnography (LL2: 273–286). In heated exchanges in the 1990s first with Collins and Yearley (Collins and Yearley 1992; Callon and Latour 1992; De Vries 1995) and later with Bloor (Bloor 1999a; 1999b; Latour 1999a), the gap between Latour and the social studies of science community widened.

To mark the contrast, Latour decided to adopt 'actor-network theory', a name suggested by Michel Callon and John Law, as the battle cry for his approach to follow and to describe the work that goes into making reality – what *is* – visible. To bluntly call his work 'philosophy' was apparently still a bridge too far. In later years he would deeply regret this decision since it utterly misnamed what the approach is up to: "There are four things that do not work with actor-network theory; the word actor, the word network, the word theory and the hyphen!" (Latour 1999b). Eventually, however, he decided to accept the name because it is "so awkward, so confus-ing, so meaningless that it deserves to be kept" (RAS: 9).

In 1982, Latour had joined economist and sociologist of technol-ogy Michel Callon at the *Centre de Sociologie de l'Innovation* (CSI) of the *École Nationale Supérieure des Mines* in Paris, one of France's top engineering schools, for what would become a fruitful co-operation lasting almost 25 years. With no administrative duties and a relatively low teaching load – one class of engineering students on Fridays and a biweekly seminar for CSI's PhD students – Latour could devote his full energies to writing about scientific practice, technology, and eventually a much wider variety of subjects. His rapidly growing international fame in science studies led to visiting professorships at the University of California at San Diego, Harvard's History of Science department, and the London School of Economics, to several honorary doctorates, and to invitations to teach in universities from Melbourne to São Paulo, all over Europe and in the US. In the 1990s, Latour started to apply the approach and the lessons learned in science studies to other subjects: art, law, politics, religion, and the ecological problems the world faces. In 1991, arguing that we fundamentally misunderstand the modern condition we live in, he published a radical and to many readers puzzling philosophical essay, *Nous n'avons jamais été modernes* – We have never been modern (NBM).

For a long time, Latour was more famous outside France than in Paris. Being at odds both with main currents in French philosophy and with French social science, and his work for a long time being almost exclusively devoted to science and technology – a subject most French intellectuals take little interest in – it took quite some time before the excellence of his work was recognized in France. Only a few heterodox French scholars – including Boltanski, Thévenot, and Descola – incorporated some of Latour's ideas in their own work. This gradually changed in the early 2000s. In 2006, Latour left the *École Nationale Supérieures des Mines* to become professor at *Sciences Po*, a prestigious (private) political and social science university in Paris, where he was elected vice-president of research a few months later. A weeklong meeting at Cerisy-la-Salle in 2007, on the occasion of Latour's sixtieth birthday, may mark his full acceptance in France as a philosopher. In rooms lined with photos of earlier Cerisy-meetings with towering figures of continental philosophy – including Heidegger, Sartre, Bachelard, and Foucault – about 100 people gathered to discuss with Latour the draft of a new work. The topic: 'exercises in empirical metaphysics'. The ambition: to summarize the enquiry Latour had been pursuing for decades and to answer the question positively of what characterizes the current situation in the West, that is, to answer the question that baffled many readers of *We Have Never Been Modern*: if we are 'not modern', what makes the civilization that brought us – among other things – the institutions of science, modern law and democracy stand out? Published in French in 2012 and in English a year later (AIME), the book was well received in France.

In 2013, Latour was awarded the prestigious Holberg memorial prize for his ambitious analysis and reinterpretation of modernity, and for challenging modernity's fundamental concepts. As the Holberg Prize academic committee noted, by then "[h]is influence has been felt internationally and well beyond the social study of science, in history, art history, philosophy, anthropology, geography, theology, literature and legal studies".

1.3 The power of addition

With hindsight, we may find that Latour did not come completely unprepared to Ivory Coast to enter on the course that would lead him to do philosophy with ethnographic means and to get engaged in 'exercises in empirical metaphysics'.

In one of his rare autobiographical writings, Latour (2010a) recounts that at the University of Dijon he had the good luck of befriending André Malet, a former Catholic priest who had become a university professor and Protestant pastor. Under Malet's guidance, Latour discovered biblical exegesis. Malet had just finished the French translation of Rudolf Bultmann's *Die Geschichte der synoptischen Tradition*, originally published in 1921, but still one of the standard texts of New Testament exegesis.

To enable modern readers to appreciate the Gospel, Bultmann, a Lutherian theologian, had decided to de-mystify the New Testament by close reading, using a method he called 'form-criticism'. Form-criticism aims to identify secondary additions to determine the original form of a piece of narrative, a dominical saying, or a parable. By eliminating step-by-step the elements and language later interlocutors had added, Bultmann eventually identified a limited number of sentences in Aramaic that could be genuinely attributed to a certain 'Joshua of Nazareth', the historical Jesus. By eliminating the later mythical additions, Bultmann hoped to open up the Gospel to modern, rationalized readers, wary of abstruse elements.

Bultmann introduced the notion of an evolving network of connections to Latour. Latour, however, suggested an alternative reading of the corpus. While Bultmann had proceeded by *subtracting* everything that couldn't genuinely attributed to the historical Jesus, Latour decided to emphasize the importance of the links that had been *added*. This led him to suggest the exact opposite of Bultmann's conclusion, namely that "only in the long chain of continuous inventions [by later interlocutors] the *truth conditions* of the Gospel resided – provided that is, that those inventions were done, so to speak, in the *right key*" (Latour 2010a: 600). In a way, he commented in 2010, "I had taken the poison out of Bultmann and transformed his critical acid into the best proof we had that it is possible to obtain truth (religious truth, that is) through an immense number of mediations provided that each link was renewing the message in the 'right manner'." The question that remained to be answered was how to define this 'right manner' precisely enough, that is "how to discriminate between two opposite types of betrayal – betrayal by mere repetition and the absence of innovation, and betrayal by too many innovations and the loss of the initial intent" (Latour 2010a: 600).

Close reading of Péguy's *Clio*, "the topic and manner of which was precisely on the question of good and bad repetition" (Latour 2010a: 600), suggested an answer. In his 1975 *thèse* and his subsequent 1977 paper 'Why does Péguy repeat himself? Is Péguy unreadable?', Latour analysed Péguy's "unceasing digressions, these monstrous paragraphs, these violent accelerations" (Latour 1977: 79), indeed Péguy's "illegibility", as exerting a calculated effect. "Repetition is the engine of war invented by Péguy to combat refrain and nagging" (Latour 1977: 80), to force the reader to break away from his habit of 'horizontally' reading a text. Persuaded to read the text over and over again, a new, 'vertical' dimension emerges. "Repetition cajoles being into time, whereas refrain crushes time in being" (Latour 1977: 80). Péguy's *Clio* does not speak about things, but about their movement, not about phenomena, but about what creates them. "In this work temporality personified [by Clio, in Greek mythology the muse of history] speaks about temporality and about a great many other things, which do not appear to be connected except by the association of words, but are connected nevertheless because she [Clio] *applies that of which she speaks*. Thus, through an incredible concentration the structure, the theme, and the style coincide to reveal the machinery of time" (Latour 1977: 81). Péguy's style, Latour concluded, has a non-stylistic basis. This led Latour to subsequently analyse Péguy's much discussed Catholicism. Again his style, rather than his use of religious concepts, is emphasized. "With Péguy's repetition, each person hears the Word in their own language; *repetition renews the work of Whitsun"* (Latour 1977: 93). For Latour, Péguy is not a prophet in the banal sense of the word, not someone who refers to God up there or to the Second Coming of Christ to talk *about* the Gospel, but nothing less than a new evangelist, "the one who brings the Glad Tidings of the loss of former dispositions, the good news of past events, of the perennial openness of the Event" (Latour 1977: 94).

It is tempting to think of Latour's encounter with the work of Bultmann in his years as a young student and his subsequent analysis of Péguy's *Clio* as unveiling the origin of his later work, including his work in science studies. In fact, there are good reasons to do so. In *Laboratory Life*, Bultmann is explicitly mentioned (LL: 169, 188 n.10) and Schmidgen (2011; 2013) has argued that a technique similar to the one used to analyse *Clio* has guided Latour's analysis of what goes on in a laboratory. Also in his later works, Latour would time and again return to Péguy (PF: 51; IRR: 1.2.6; REJ: 72; AIME: 249, 306; Latour and Howles 2015).

As he observed, looking back in 2013,

> [w]hat is certain is that I emerged from that formative period armed with total but somewhat paradoxical confidence about the fact that the more a layer of texts is interpreted, transformed, taken up anew, stitched back together, replayed, and rewoven, each time in a different way, the more likely it is to manifest the truth it contains – on condition (this is the part I retained for later use) that one knows how to distinguish it from a different mode of truth, pure and perfect information [. . .]. A long struggle against the eradication of mediations was about to begin. (Latour 2013b: 289)

Latour had detected what he would fight against for all his career: the idea that information flows effortlessly, that truth does not require work – an idea he would polemically call in his later work (e.g. AIME: 93), alluding to the computer mouse, 'double click' communication.

However, the temptation to see all of Latour's later work as just an extension of his early work in biblical exegesis and his theological interpretation of Péguy should be handled with care. When Latour claims that religious truth is obtained through an immense number of mediations rather than residing in the few sentences that Bultmann succeeded to attribute to the historical Jesus, it would be quite paradoxical to claim that the true message of Latour's work resides in ideas he expressed as a young man. If we want to understand Latour's philosophy, we too have to follow his movements, rather than to try reconstructing the 'origins' of his work. It would be superficial to think that Latour's 'empirical philosophy' is biblical exegesis writ large, or to locate its origin in Latour's moment of epiphany on the road from Dijon to Gray. His 'empirical philosophy' took shape and substance in his moves, in ethnographical studies, in the debates he got involved in, and in the way he incorporated ideas from a wide variety of sources in his work. Empirical philosophy, too, exists because of the many *links* that have been assembled; not because some idea sparked in the mind of an *agregé de philosophie* in his early twenties.

The usual trick when interpreting a philosopher of citing 'origins' and 'context' should therefore be resisted. What makes Latour's work stand out is his *style* of doing 'empirical philosophy'. Like Péguy's Clio, Latour "applies that of which he speaks". Concepts are introduced as tools and discarded when more useful ones are found; precursors in philosophy and social science are pragmatically introduced to help convey the message. When writers with better ideas and concepts are found, former ones are kindly invited to leave the stage. There are numerous references to other thinkers in Latour's writings – to Austin, Deleuze, Dewey, James, Kant, Nietzsche, Serres, Sloterdijk, Souriau, Spinoza, Stengers and Whitehead, to Foucault, Habermas and Heidegger and to non-philosophers like Braudel, Garfinkel, Greimas, Tarde, Tolstoy and Tournier – but the usual intellectual stratagem of tracing influences and spotting differences and resemblances is of little use most of the time. When Latour conducts detailed exegesis of former thinkers and writers (e.g. of Serres, Souriau, Whitehead and Tarde), he discusses them to make his own points. The ultimate test is whether their work and ideas help to enrich ethnographic field-work and to let us better see what we see. Due to the movement of his work, the delight of minutiae it exposes, and the light-heart-edness of the style of most of his writings, like Hemingway's Paris, Latour's philosophy is 'a moveable feast'. To get acquainted with it, we better start to follow his moves.

2

Science Studies

For most of his career, Latour considered himself to be primarily an 'anthropologist of science and technology', as someone engaged in 'science studies'. Science may not have been his first love, but the chance meeting in Guillemin's Salk Institute with the practice of science was the start of a life-long love affair.

The affair made science's other suitors nervous. In spite of Latour's explicit denials, his work was perceived as a specimen of postmodernism and as a threat to the calling of science. He was accused of undermining "the vision of a rationally understandable world" that "protect[s] ourselves from the irrational tendencies that still beset humanity" (Weinberg 1996: 15; cf. Gross and Levitt 1994). Surprisingly, in due course, his work incited also severe opposition from his colleagues in science studies. Here, he was not charged for undermining rationality, but for basically the opposite offence, namely for having "nothing to offer on the question of the primacy of human society in the making of knowledge [. . .]" (Collins 1994: 674), and for developing an "essentially conservative [. . .] prosaic view of science and technology" (Collins and Yearley 1992: 323).

Although framed in political terms, at stake are philosophical issues about the nature of science and the cognitive authority attributed to science in modern societies, and ultimately about the relation of man and the world. Redescribing the practice of science, Latour distanced himself from intuitions and conceptual distinctions that the philosophical tradition has taught us to hold dear. It aroused disbelief and was conceived as a "scholarly joke about the world and its scientists" (Collins 1994: 672). "Surely, you are joking, monsieur Latour!", a baffled reviewer of *Science in Action* exclaimed (Amsterdamska 1990). They were wrong.

In order to understand these reactions and to see why they missed the mark, let's first review how philosophers have accounted for science. They addressed two questions. Firstly, *what is science?* – that is, how does science proceed, what are its methods, how is scientific knowledge produced? And secondly, why do we attribute more authority to scientific knowledge than to other forms of belief? – that is, *what's so great about science?* (Feyerabend 1976: 110). Focusing on the special nature and authority of scientific knowledge, philosophers discussed science in *epistemological* terms.

For a long time, philosophers assumed the answer to the second question – what's so great about science? – to be obvious. Science is great because it succeeds in attaining – or at least in approximating – truths about nature, by representing the 'real world', the 'world of our experience'. This implies also the answer to the first question: science proceeds by generalizing careful, unprejudiced observations, which provide both the foundations on which scientific knowledge is built and its final court of appeal.

At the beginning of the twentieth century it was, however, clear that these answers were too simple. The most advanced discipline of the time – physics – had seen major advances, both theoretically and experimentally. The school that dominated philosophy of science for the first half of the twentieth century, logical empiricism, therefore decided to limit itself to the study of the products of science, to establish under which conditions the claim can be logically justified that a scientific theory has 'empirical content', that is, relates to observation statements. The process of creating knowledge, the 'context of discovery', was left as a subject to future psychologists. The proper domain for philosophy of science was limited to the 'context of justification'.

Even in this restricted form, the task still presented a tenacious challenge. Theories of physics can be formulated in various – mathematically equivalent – forms that present radically different pictures of the real world. For example, at the end of the nineteenth century, Hertz, dissatisfied with the obscurity of the concept of 'force' in Newtonian mechanics, had shown that it is possible to formulate classical particle mechanics without using the concept of 'force', thus giving a representation of the physical world that is quite different from the usual one. So, the question became how to distinguish between on the one hand the pure empirical content of a theory and on the other hand its conventional elements, introduced to present the empirical content in a suitable form.

Conceiving scientific theories as sets of statements and declaring that the meaning of a statement is its method of verification suggested that logical analysis might provide the answer. The logical empiricists set out to decompose theories – which they conceived as sets of propositions – along two dimensions, *viz.* the 'conventional dimension' (comprising linguistic elements, the concepts, mathematics and other conventions that scientists have chosen to formulate and to communicate their ideas, making up the parts of theories that are true by virtue of meanings and independently of facts) and the 'empirical dimension' (the 'empirical content', the part of the theory that allows empirical verification). Once the decomposition had been completed, it would also be possible to more precisely answer the first question about science – how does science proceed – by setting out what scientists should aim at, namely theories with maximum empirical content, and how to choose between competing theories, *viz.* on the basis of the 'degree of confirmation' of their empirical content.

However, in spite of this *prima facie* straightforward approach, the logical empiricist programme soon became confronted with internal problems that turned out to be solvable only by proposing increasingly intricate ideas that subsequently turned out to be ridden with internal problems of their own. Like medieval astronomers, the logical empiricists had to add ever more epicycles to save an untenable system. In 1951, logical empiricism was struck to its foundations. In a paper of only twenty-four pages, Quine (1951) showed that the whole idea that the meaning of a statement is somehow analysable into a conventional component and a factual component was an untenable "empiricist dogma", a "metaphysical article of faith". Philosophy of science had to try another tack.

Wittgenstein's (1969 [1952]) turn from logical analysis towards analysis of ordinary language practices provided the clue for how to change course. Attention was shifted away from logical analysis of the products of science to descriptive analyses of scientific practices. The corpus to work on expanded. No longer could philosophers limit themselves to studying only the established products of science, the statements one finds in textbooks. They also had to examine lab-journals, notebooks, correspondence and discussions before results had reached the state of textbook knowledge. Behind the neat order of established scientific results, they soon encountered an unruly practice, full of uncertainties, controversies and practical problems.

The turn to practice raised a number of questions. Which method should be employed to study the practice of science systematically? How are the results of rivalling approaches to the study of science to be appraised? And whose practice is worth studying anyway? Science is a normative practice – there is 'good science' and 'bad science' (or 'pseudo-science', that is, intellectual endeavours that dress up like science but don't deserve its name). The decision to study some episode in the development of science in the hope of becoming empirically informed about how science proceeds implies the implicit judgement that the people whose work is studied can be considered to be good, serious scientists. This problem was evaded by focusing on the work of scholars whose merits are beyond dispute – Copernicus, Galileo, Boyle, Newton, Planck, Einstein and the likes, the towering figures of the tradition of modern science. However, the fact that in due course even the work of such giants will become amended and in some cases even rejected, suggested that insights drawn from studying only one particular scientist would not suffice. To get a systematic view, science had to be studied comparatively and in its development as a practice that involves both tradition and innovation.

This, however, revealed a deeper problem. It turned out that what makes the towering figures in the scientific tradition stand out were not only their empirical results, but also the fact that their work involved radical innovations in method and style. The formerly widely held idea that there is one unique set of rules, 'the scientific method', that characterizes good scientific practice, turned out to be an illusion. "Given any rule of method conceived as 'fundamental' or 'necessary' for science, there are always circumstances when it is advisable not only to ignore the rule, but to adopt its opposite," Feyerabend concluded on the basis of a review of various historical episodes; "the only principle that does not inhibit scientific progress is 'anything goes'" (Feyerabend 1975: 23, 28).

Kuhn (1970) had already been more specific. *The Structure of Scientific Revolutions* showed that the very idea that in science theory-choice is made on the basis of methodological rules is fundamentally flawed. Such choices presuppose agreement among scientists not only about standards of appraisal, but also about which problems are worthy to be tackled. However, Kuhn's historical studies showed that in the development of science phases where this kind of agreement indeed exists are interspersed by phases characterized by its absence.

In the first phase, which Kuhn called 'normal science', scientists work on a limited set of puzzles suggested by a 'paradigm', the 'disciplinary matrix' of their specialty, a framework consisting of the symbolic generalizations, beliefs in particular models, and the values and exemplars scientists of that particular epoch have been trained to follow. In this phase, when a scientist fails to solve the problem he is working on, it is his competence that is questioned, rather than the adequacy of the disciplinary matrix and the theories that are built within this framework. Only when problems that resist even the efforts of the most experienced scientists multiply, does a 'crisis' emerge and do discussions about the fundamentals of the disciplinary matrix arise. However, in this second phase, in episodes of 'crisis', the agreement necessary to evaluate the relative merits of competing theories is characteristically absent. "Each group uses its own paradigm to argue in that paradigm's defense" (Kuhn 1970: 94). So, Kuhn observed, the historical record suggests that even when it is possible to test a theory (in the phase of 'normal science'), it is hardly ever done; whereas during a crisis, when fundamental choices are really at stake, a test is not possible. "The choice [. . .] between competing paradigms proves to be a choice between competing modes of community life" (Kuhn 1970: 94). This led Kuhn (1970: 210) to conclude: "Scientific knowledge, like language, is intrinsically the common property of a group or else nothing at all. To understand it we shall need to know the special characteristics of the groups that create and use it."

The characteristics of scientific communities are the proper subject of sociology of science. Before Kuhn, sociologists had limited themselves to studying the social conditions of science as a practice, such as its institutional base and normative structure, its reward system, and age structure (e.g. Merton 1973). Questions about scientific knowledge had been left to scientists themselves; while questions about the epistemological justification of knowledge claims were entrusted to philosophers. *The Structure of Scientific Revolutions* opened the gate for extending the domain of sociology of science to engage in 'sociology of scientific knowledge', to explain the content of science in sociological terms.

Following Kuhn's lead, science studies took 'a social turn'. To explicitly distance empirical sociology of knowledge from philosophical attempts to account for science, Bloor (1976: 4–5) formulated four leading methodological principles for what became known as the 'Strong Programme in the Sociology of Knowledge'. To answer the two key questions about science – how does science proceed and why do we attribute authority to its results – the sociology of knowledge would be:

(1) *causal*, that is, concerned with the conditions which bring about belief or states of knowledge;
(2) *impartial* with respect to truth and falsity, rationality or irrationality, success or failure. Both sides of these dichotomies require explanation;
(3) *symmetrical* in style of explanation. The same types of causes would explain, say, true and false beliefs;
(4) and *reflexive*. Its patterns of explanation would have to be applicable to sociology itself.

With these methodological principles a lively and ambitious new field, 'social studies of science', was launched.

2.1   The 'Sociology of Scientific Knowledge'

Social studies of science developed in the 1980s in various variants. The 'Strong Programme' focused on the role of 'external' factors – professional interests and political ideologies – in the development of scientific knowledge. Other sociologists studied the social processes within scientific communities. Collins, who later became one of Latour's most outspoken critics, is one of them. Collins called his approach 'the Empirical Program of Relativism' and later renamed it 'Sociology of Scientific Knowledge'. To assess the philosophical divide between Latour and Collins, the latter's *Changing Order* (Collins 1985) needs to be discussed in detail.

Changing Order sets out the intellectual underpinnings and empirical ambitions of Collins' 'Sociology of Scientific Knowledge' and shows in detail how epistemological questions are translated into sociological ones. Its point of departure is a simple observation: although the world comes to us at birth in what William James (1950 [1890]: 488) once described as "one great blooming, buzzing confusion", there is concerted perception and understanding in science. Collins (1985: 5, 6) subsequently conceived the task of the 'Sociology of Scientific Knowledge' as analysing *how* such concerted perception and action come about, that is, how scientists come to be certain about regularities. As such, Collins' problem is a well-known, traditional philosophical issue, *viz.* the problem of induction. However, Collins' ambition was to approach this problem by empirical means. To get there, a few steps are necessary.

To begin with, Collins decided to work on a particular version of the problem of induction, known as 'the New Riddle of Induction' (Goodman 1973), which translates the epistemological problem into a problem of language-use. What makes us expect that we can apply the same words and phrases to the same things and events the next time we encounter these things and events?

Wittgenstein had provided an answer: we are following a rule. Collins follows Winch's (1958: 32) exposition of Wittgenstein's idea: "The notion of following a rule is logically inseparable from the notion of making a mistake." How do we decide whether we have correctly followed a rule, that is, that we used the correct word or phrase to speak about a thing or event? And if we have used the wrong word to refer to a phenomenon or thing, the phenomenon or thing itself will not protest. But if we have made a mistake, other people will be able to correct us. Hence, Winch concluded, meaning is constituted by a community providing agreement. That agreement is not an agreement of opinion but one of being socialized in a 'form of life'. Collins translates this conclusion in sociological terms: "[i]f cultures differ in their perception of the world, then their perceptions and usages cannot be fully explained by reference to what the world is really like" (Collins 1985: 16). Coordinated perception, correct language use and production of knowledge are all based on being socialized in a community.

At this point, all the necessary connections for launching the 'Sociology of Scientific Knowledge' are in place. What philosophers have discussed in epistemological terms as the problem of induction is turned into a problem of socialization, a problem that sociologists, the experts on community life and socialization, can study.

So is the usual image of science, articulated by philosophers and scientists alike, a complete myth? Do methodological rules and observation play no role at all in science? Well, in fact they do. But, as Collins (1985) states, they cannot *fully* explain theory-choice and other aspects of scientific development because both rules and observations are subject to 'interpretative flexibility'. Rules do not contain the rules for their application. To be able to correctly follow a rule requires being socialized in a form of life. That is what closure on debate is based on.

To flesh this out, Collins studied the role of an uncontroversial rule of empirical science, the demand of reproducibility. To experiment is to produce, refine and stabilize phenomena; an experiment that produces phenomena that cannot be replicated by anyone who has learned the relevant techniques has failed to do its job. Only results that can be replicated count as objective facts. However, what *counts as* a replication of an observation or an experiment? To answer this question, Collins decided to study the process of reaching agreement in science.

One of Collins' case studies involved the claim of Joseph Weber, a physicist of the University of Maryland, to have detected gravity waves. The existence of gravity waves had been predicted by Einstein's theory of relativity. In principle a flux of gravity waves might be observed as an effect of major stellar events (such as exploding supernovae, or stars collapsing in a black holes) on g, the gravitational constant on Earth. To detect these waves, Weber had built an enormous 'antenna', a bar of aluminium alloy of several tonnes housed in a vacuum chamber. The idea was to measure changes in the length of this bar that were caused by the energy pulses of gravity waves. Because the expected effects were extremely small, precautions had to be taken to eliminate or correct for intervening effects, e.g. seismic vibrations and magnetic and electric disturbances. In 1969 Weber claimed to have detected about seven peaks every day that could not be accounted for by noise. A few years later, he found that signals could be detected simultaneously on detectors separated by thousands of miles. He also found that there was a periodicity in the disturbances of around twenty-four hours. This was Weber's *eureka* moment: his findings suggested that the fluxes of waves that were registered by his detectors came from extra-terrestrial sources. However, the effects Weber claimed to have measured were orders of magnitude larger than predicted by Einstein's theory. If Weber was right, fundamental theories in physics and cosmology had to be revised.

Several laboratories therefore started to build their own antennae to detect gravity waves. They all reported negative results; the effects Weber claimed to have detected could not be replicated. So had Weber been wrong? Weber denied this; he claimed that his competitors had not used the right equipment and had used inadequate procedures for distinguishing signals and noise. A controversy emerged. About this time, Collins started to interview the physicists involved.

Had Weber's claim indeed been refuted by the other laboratories? Or do gravity waves hit the Earth and did the other laboratories fail to detect them because they used the wrong equipment? There is only one way to find out: we too must build a good wave detector and look again. But what is a 'good' detector? We won't know until we have tried it and found the correct outcome! But we won't know what the correct outcome is until . . . . and so on *ad infinitum*. We are stuck in a vicious circle, which Collins called the 'experimenters' regress'. We have to find a way to break into this circle. How to decide what is a good detector?

Interviewing the physicists involved, Collins noted that although nearly all agreed that Weber had failed to detect gravity waves, their reasons differed markedly. Technical issues and mistakes that some physicists considered as being of utmost importance, other interviewees viewed as irrelevant details. Moreover, many interviewees also criticized the experiments of colleagues that had reached negative results. Collins found that in discussing Weber's work, his critics also provided comments on social matters and on issues of competence. For example, it was observed that Weber was not employed by one of the major universities, that he had little experience with large-scale physical research, had few contacts in the physics community, and had a background in engineering rather than in physics. Some even questioned his personality. Such judgements, Collins concluded, should not be discarded as 'mere gossip', because they have an epistemological effect. They suggest that a convincing experiment is an experiment conducted by someone who is conceived by his peers to be competent and well-socialized.

Collins (1985: 89) concluded:

> the definition of what counts as a good gravity wave detector, and the resolution of the question whether gravity waves exist, are congruent social processes. They are the social embodiment of the experimenters' regress. [. . .] When the normal criterion – successful outcome [i.e. replicable results] – is not available, scientists disagree about which experiments are competently done. Where there is disagreement about what counts as a competently performed experiment, the ensuing debate is coextensive with the debate about what the proper outcome of the experiment is. The closure of debate about the meaning of competence is the 'discovery' or 'non-discovery' of a new phenomenon.

To summarize, Collins claims: 1. Scientific facts are established only after closure of a discussion in a scientific community;

2. In the community, outcomes and technical details about the experimental set-ups and procedures are discussed; 3. When disagreement continues, closure is brought about by judging the competence of the experimenter – a judgement that typically is based on a wide variety of considerations, including considerations about the degree to which the experimenter is conceived as properly socialized. So what science textbooks contain are not statements of 'brute facts', but statements of 'institutionalized facts', that is, facts that depend for their existence on human institutions, beliefs and social processes. In other words, scientific facts are 'socially constructed'. So "[i]t is not the regularity of the world that imposes itself on our senses but the regularity of our institutionalized beliefs that imposes itself on our world" (Collins 1985: 148).

How much progress had been made? Philosophically speaking, not much. Kant (1956 [1787]) already argued that we have no access to 'things-in-themselves' – Kant's version of 'brute facts'. The world we perceive is what Kant called the phenomenal world, not the noumenal world. Where Kant pointed to the synthetic role of the pure forms of intuition and the categories of the mind in perception and understanding, Collins refers to socialization and the regularity of institutionalized beliefs. In fact, Kant has the better hand, because Collins' solution – referring to the *regularity* of institutionalized beliefs – introduces the problem of induction at a new level. Moreover, for Kant, apart from the faculty of understanding, human beings are also endowed with Reason – capital R. While humans cannot perceive 'things-in-themselves', they have *reason* to think that they are on the other side of appearance: "even if we cannot *cognize* [. . .] objects as things-in-themselves, we at least must be able to *think* them as things-in-themselves. For otherwise there would follow the absurd proposition that there is an appearance without anything that appears" (Kant 1956 [1787]: B xxvi–xxvii). Replacing Kant's transcendental Ego by institutionalized beliefs, the position Collins defended is a sociological, reduced version of Kantian epistemology.

Empirically, however, by opening up knowledge-production for sociological investigation, Collins' move shows decisive advantages. His case studies provide details about the practice of science that predecessors – philosophers, sociologists and most historians of science alike – had left untouched. No one had doubted that science is a human affair and that reaching agreement plays a crucial role, but Collins provided the details.

Still, many questions remained unanswered. Aren't there *specific* social processes in play – practices, procedures that one may find in science, but nowhere else? What role do instruments play in the development of science? If scientific knowledge, emerging out of social processes, is merely a 'social construction', why did both Weber and his competitors take the trouble to drag tons of aluminium alloy or other materials to their laboratories and to use computer programs to perform complicated statistical analyses? Has science not become a bit all too human in the hands of the sociologist of scientific knowledge?

To answer these questions, we need to visit a laboratory, the place where the production of scientific knowledge-claims takes place. What are people *doing* in a lab apart from discussing? Latour will be our guide.

2.2   An anthropologist visits a laboratory

When Latour went to California in 1975 to study for almost two years the daily activities of scientists and technicians in the San Diego Salk Institute, his interests came straight out of his earlier work in Ivory Coast that led him to question prevailing views on cognitive competences. Only after the chance meeting with Steve Woolgar at a conference in Berkeley in the summer of 1976, did Latour discover that he shared an interest with people involved in the emerging discipline of 'social studies of science'.

At the time of Latour's visit, the Salk Institute in San Diego devoted its efforts to extracting, isolating and determining the chemical structure of substances that were supposed to regulate a variety of hypothalamic hormonal processes. The central problems of the research were technical, not conceptual ones. The relevant (peptide) chemistry was well known; the main problem the researchers had to face emerged from the improbably small amounts of these substances in human (and other mammalian) brains. The research therefore concentrated on trying to extract small amounts of these substances from literally tons of sheep brains and to determine their chemical structure as accurately as possible. For that purpose, the Salk Institute experimented also with synthetically produced compounds to detect which of them had similar biological effects. Shortly after Latour had left the Salk Institute, its director, Guillemin, was awarded, with Schally, the 1977 Nobel Prize in Physiology or Medicine for their discoveries concerning the peptide hormone production of the brain. That the Salk Institute was involved in 'good science' is beyond dispute.

The people at Salk are neuroendocrinologists. They considered themselves more generally speaking to be 'natural scientists'. However, Latour soon observed that this was quite a puzzling denomination. Apart from a few pot plants in the secretaries' offices and some caged rats and guinea pigs, there was no 'nature' to be found in the laboratory at all. At first sight, the Salk laboratory had the appearance of a small factory with a top-heavy administrative department. Those who worked there were focusing their attention on apparatuses and on papers.

Latour found the laboratory to be separated in two sections. One section, the domain of technicians and analysts, contained various items of equipment. The individuals in this section were engaged in mixing chemicals, shaking Erlenmeyer flasks, setting up assays, handling electronic equipment, and preparing animals for experiments. When the people who work here were not handling pieces of equipment or laboratory animals, they were spending their time writing numbers on the side of tubes or on the fur of rats, filling in large books with long lists of figures to record what they had just done, or drawing up tables and graphs. The other section housed the scientific staff. Their offices were populated with texts of various sorts: graphs, tables, computer printouts, handbooks, scientific journals, preprints of articles, conference reports and so on. Most of the time, the scientific staff was engaged in talking, reading and writing. By juxtaposing various types of texts, they produced new tables, new graphs, new papers, and new handbook chapters. They were engaged in proliferating texts.

How is it possible that in a community of people whose main interest is focused on coding, marking, altering, correcting, reading and writing, the idea has emerged that they are studying 'nature', the 'real world', the 'world of our experience'? To solve this riddle, Latour studied the way in which, both within the laboratory and in the scientific community at large, texts circulate and become transformed, and how in this process facts are constructed.

First, he observed the importance of *inscription devices*, a term Latour introduced for any device that directly relates a material substance (like chemicals, lab animals, etc.) to a figure or diagram that is usable by one of the scientists in the second section of the lab. The meticulous marking and coding by the technicians and analysts served this purpose. In other cases the intervention of technicians was circumvented, as some equipment may produce e.g. automated chemical analyses to report results directly in the form of graphs or data sheets. The beauty of inscriptions is their mobility, stability and combinability. Once an inscription is available, it can be easily transported, stored, and combined or compared with other inscriptions. They may travel through time and space more reliably than human memory and they are much less subject to degradation than chemicals and animals. They are what Latour later called *immutable and combinable mobiles* (SA: 227, 236). Inscriptions *translate* the conditions of an experiment and its outcomes (for example the animal used, the composition of its blood before and after a compound has been injected, its behaviour, etc.) into text or figures. Once inscription devices have translated chemicals and the behaviour of lab-animals into figures or text, there is no need to bring the substances and animals into the scientists' offices. The scientists can focus their attention on the figures, texts and graphs that are handed in by the lab technicians. They provide the 'data' that wait for further analysis. The scientists can compare them with other figures and graphs, combine them, analyse them, insert them into their own notebooks, or in a document that may eventually grow into a scientific paper. However, if necessary, for example when a human error or malfunction of apparatus is suspected, it is possible to go back to the original substance (safely stored in a refrigerator, with a code on the tube).

So, on the desks of the scientists, there are only piles of texts: journal articles, handbooks, data sheets, graphs, tables, and computer printouts. Some of these texts have been delivered by mail, some have been produced automatically by machines, others have been brought by lab technicians. Before this heterogeneous set of texts has arrived on his desk, a scientist will have read books and journals, made notes of his own, discussed issues with his colleagues, and written instructions for the technicians to follow when setting up an experiment, the results of which are now waiting on his desk. On the basis of these texts, by weaving the various elements into a dense argumentative network, the scientist will compose a new text. In the new document that is prepared, the inscriptions, figures and texts that are used will get a new meaning. For example, what a predecessor has reported as a final conclusion, may now function as a cornerstone for a new argument, or may be reported as being inconsistent with the results of the experiments performed by the lab technicians. In the document the scientist produces, the texts that are used to write the new document are *translated* again in both senses of the word: they are moved from one place to another and in this process they get a new meaning. They are *enrolled* to make a claim – e.g. the statement that a rat injected with a substance with chemical structure XYZ shows an enhanced form of behaviour ABC, suggesting that this type of behaviour in mammals is induced in the mammalian brain once XYZ has been injected.

If the scientist succeeds in convincing a journal's referees and its editor that his paper is worthy of publication, the fate of his statements lies in the hands of its readers. If no one takes the trouble to read the publication, his efforts and the work of the technicians and analysts who supported him will have been in vain. But suppose someone does take the time to read and comment on the article, for example after comparing its claim with claims made by other scholars, or with the results of the reader's own attempts to reproduce the experiments. The reader's findings may suggest that the statement needs modifications. For example, he may report that his experiments show that a chemical with structure XWZ rather than XYZ induces an even more convincing ABC behavioural pattern in rats; or that XYZ induces ABD rather than ABC patterns of behaviour; or he may observe that the claim that 'the natural form of this type of behaviour in mammals is probably controlled by a substance with XYZ chemical structure' is still unfounded because it is still unknown whether XYZ will have the same effects in other mammals. In short, subsequent readers will add *modalities* to the original statement that qualify the original claim. Rather than a factual claim, it becomes 'an unfounded suggestion', or 'a promising hypothesis that however needs to be corrected in details', etc.

In this process, the statement and its original author have become separated. The statement that circulates acquires a 'career' of its own. In due course, it will become modified, and to its later versions new modalities will be attached. This process may continue for quite some time. If modalities continue to be added, and doubts proliferate, scientists will likely turn their attention to more promising research topics. The original statement and all the attempts to present it in a form that is more defendable will fade into oblivion. However, if the process reaches a point where all modalities are dropped, a *fact* has been established. It is accepted that 'a mammal injected with a substance with chemical structure XYZ will show an enhanced form of ABC behaviour'. At this point, an important *inversion* has taken place. All earlier doubts and unfruitful detours will be forgotten, and history will be rewritten by inverting process and outcome. From this point on, the process will be narrated as the search for this particular outcome: this is the fact that the scientists have been looking for. It took quite some effort before it was discovered, but by now its existence has been revealed. From now on, the history is perceived from this vantage point: the process of scientific discovery is turned into the pursuit of a single path which inevitably led to the discovery of the structure of this substance, and its role in mammals. All uncertainties and side-paths, all traces of the many modalities that may have been attached in the course of time, will be discarded. The scientist who has launched the original statement will be celebrated as the one who was right and who has seen further, deeper and more accurately than anyone else. In an autobiographical statement, he may recall all the doubts that had to be overcome. But all of that is now history, not science.

Facts are constructed out of texts by this *process of splitting and inversion*, Latour and Woolgar claim. The remarkable correspondence between statement and fact that is called truth and that has puzzled philosophers for ages evaporates. By focusing on the *process* in which this correspondence becomes established, it becomes clear that this correspondence is not established by a unique human ability to cross the abyss between language and reality, nor enabled by some 'logical form' that Wittgenstein's (1969 [1921]) *Tractatus* suggested is shared by language and reality. To achieve correspondence of statements and facts requires a long process of scientific *work*, of establishing many *translations*, in which *both knowledge and reality* are transformed. To achieve correspondence, scientists not only had to reach agreement, but also had to ask the lab-technicians to inject rats with substances they had carefully prepared, to set up assays, to register their behaviour. The 'process of splitting and inversion' hides this *double* process from view, leaving us with the idea that brilliant scientists have been able to make miraculous jumps to discover facts that have been waiting for ages to be discovered.

When Latour and Woolgar summarize their analysis by noting that "the work of the laboratory can be understood in terms of the continual generation of a variety of documents, which are used to effect the transformation of the statement types and so enhance or detract from their fact-like status" (LL: 151), we may recognize the crucial lessons from Latour's encounters with Bultmann and Péguy in his student days. In the first place, the emphasis is put not on statements, but on their movement, on the way in which statements are transformed and how in this movement facts are created. Secondly, it is recognized that once this is done the truth conditions reside in the enhancing and detracting of the fact-like status of statements – that is, in a subtle balance of repetition and innovation. Thirdly, instead of dealing with this issue in the usual way, that is, in terms of 'logic' and 'reasoning', Latour and Woolgar analyse the matter by focusing on "the routine exchanges and gestures which pass between scientists" (LL: 151), that is, their style. While Latour found the conditions of religious truth in the way Péguy arranges words and phrases, Latour and Woolgar claim that the conditions of scientific truths are located in the way scientists arrange inscriptions, statements and documents into a network comprising long chains of translations. It would take Latour twenty years to specify the proper conditions for the style of science:

> an essential property of [these] chain[s] is that [they] must remain *reversible*. The succession of stages must be traceable, allowing to travel in both directions. If the chain is interrupted at any point, it ceases to transport truth – ceases, that is, to produce, to construct, to trace and to conduct it. *The word 'reference' designates the quality of the chain in its entirety*, and no longer *adequatio rei et intellectus*. (PH: 69; cf. also Chapter 6)

When *Laboratory Life* was published in 1979, most readers understood Latour and Woolgar's description of the process in which a scientific fact is constructed out of texts as another contribution to 'social studies of science'. To them, Latour and Woolgar seemed to suggest that scientific knowledge is produced in 'social processes' and that facts are 'socially constructed'. Both the original subtitle of *Laboratory Life* – 'the social construction of scientific facts' – and the philosophical tradition that, ever since Kant, associates 'construction' with the operations of human understanding, contributed to this reading. But it is plainly wrong. The mistaken reading ignores the work that goes in the handling of chemicals, equipment and lab animals – in short: all the *work* that goes on in the first section of the laboratory – and the crucial role of *inscriptions* that relate specific material items to figures, graphs, and texts. Scientific discovery not only requires (social) processes of interpretation and reaching agreement about statements; it also requires work on the other side of the long chain of translations that relate published statements to events and phenomena. To know reality, scientists have to intervene, manipulate and change reality. Doing science means being engaged in both epistemological and ontological work. To observe phenomena and to register facts, reality has first to be made visible.

2.3   Anatomy of a scientific paper

For those who don't have the opportunity to spend two years as an anthropologist in a laboratory, or to join a team of soil scientists on a field trip to Brazil (PH: 24–70), Latour showed another way to get a detailed view on the process of establishing a scientific fact (PH: 113–144; Latour 1993). Careful reading of a published scientific paper will serve too.

This claim may come as a surprise. In a BBC talk of 1963, the winner of the 1960 Nobel Prize in Physiology or Medicine Peter Medawar (1996 [1963]) famously stated that "the scientific paper is a fraud". Of course, Medawar didn't mean that scientists deliberately mislead their readers, nor that they fiddle with the facts. He meant that scientific papers misrepresent the work and thoughts that led to the results described in them. Inspired by a naive and unsustainable empiricist philosophy of science, the formats of publication that contemporary scientific journals demand from contributors cover up the process of scientific discovery.

Medawar was partly wrong. Applying an appropriate technique of analysis, we may extract from a scientific paper a lot of insight about how its results were produced. But Medawar is right that it is often difficult to apply this technique to a contemporary scientific paper. So, to learn the necessary technique, we should turn to an older publication. A paper of one of the great experimentalists of the nineteenth century, Pasteur, '*Mémoire sur la fermentation appelée lactique*' (Pasteur 1858: 404–418, partially translated in Conant and Nash 1964), perfectly serves this purpose.

Pasteur (1822–1895) became interested in fermentation – the process that turns sugars into alcohol or organic acids – in the 1850s. Although controlled fermentation had been used for ages to produce cheese and yoghurt and alcohol in beer and wine, the explanation of the process was completely obscure. Brewers at Lille, where Pasteur had been appointed professor of chemistry in 1854, asked Pasteur to look into the process. However, his interests were not merely practical ones. Major theoretical and philosophical issues were also at stake. Are living organisms fundamentally different from non-living entities; or can life be fully explained by the laws of physics and chemistry that account for inanimate nature? The school of thought called 'vitalism' claimed that in life-processes a special force was involved. Liebig, a leading German chemist, rejected this view as being obscure and unscientific.

Against the background of this controversy, fermentation provides a perfect research topic. In beer production, fermentation of sugars into alcohol requires addition of living yeast. On the other hand, lactic fermentation turns fresh milk sour in a few days without the explicit addition of living organisms. Being a chemist, we may expect Pasteur to take sides with Liebig. However, as he reported in his *Mémoire*, he was "led to an entirely different point of view" (Pasteur 1858: 408). His experiments showed that lactic fermentation also depends on the activity of living organisms. This discovery marks the birth of modern biochemistry.

Pasteur's *Mémoire* has been analysed in detail by Latour (PH: 113–144; Latour 1993a). To facilitate understanding of the technique he used for this analysis, we must first distinguish two different modes of reading a (non-fiction) text.

Usually, we will read a (non-fiction) text as a document informing us about events in the world outside. This is how we read a newspaper. Its articles report about events that have happened somewhere in the world. Suppose today's paper brings the story that some highly classified, politically sensitive material was stolen from the home of a defence ministry's civil servant. The article continues that because taking classified files home is a severe breach of security regulations, the civil servant had to report to the Secretary of State for Defence. The newspaper reports that the minister had been outraged and had fired the civil servant. Having read this story, we may raise *epistemological* questions. Is the newspaper report true, or is it a fabrication? How did the journalist know what has been discussed between the minister and his member of staff, as their meeting probably took place behind closed doors? Has the burglary in fact taken place in the civil servant's home at all? Or was the story made up by the journalist, or had it been leaked, e.g. to cover up breaches of security within the Ministry of Defence itself?

However, we may read the *same* newspaper article also in an other mode, namely *as if* it is the script of a play. The article suggests two separate scenes: a burglary and a discussion in the minister's office. Reading the article in this mode, we may raise *ontological* questions: if we want to re-enact these scenes on stage, what has there *to be* (on stage) and what kind of actions will have to be performed? Obviously, we will need a number of actors – to play the burglar, the civil servant and the minister. We will also need a few props, e.g. a file holder containing classified documents. The play will need at least two acts: one situated in the civil servant's home where the burglary takes place and one situated in the minister's office. Once all the actors and props are present, we may take our seats to watch the events unfold in the world created on stage.

Not only a newspaper story, but also a scientific paper may be read in either one of these two modes. Of course, usually we will read a scientific paper in the first mode, as a document informing us about the world outside. Undoubtedly, this is what its author expects us to do. He has written his paper to inform us about – say – the existence of the phenomenon he claims to have discovered. His text refers to the new phenomenon, and it provides the reasons for his claim that he has discovered it. However we may also read his article in the second mode, as a script for a play – to analyse what had to be present and what had to be performed to make this discovery happen and how epistemological issues were handled. This is the technique Latour uses to analyse Pasteur's *Mémoire*. He reads the *Mémoire* as if it is the script of a play. He finds Pasteur's article to be more complicated than a simple newspaper story: anticipating epistemological questions the report of his experiments will evoke, Pasteur has woven them *into* his text. Reading Pasteur's text as if it is the script of a play, Latour extracts two dramas from it: an 'epistemological drama' and an 'ontological one'.

To analyse a text by reading *as if* it is a play is basically the trick used in semiotics, a methodology – or rather a loosely structured family of methods – to analyse how meaning is produced in language, narrative texts and in cultural systems, that is, in what semioticians call a 'system of enunciation' (Greimas and Courtés 1979; Barthes 1988; Eco 1994). Latour uses a light version of semiotics, not the whole apparatus introduced by Greimas. And he boldly applies it to analyse a *scientific* article.

Semiotics is a methodology overrun by technical terms. A few of them will however be useful for our purposes. In the first place, we need to distinguish the real author of a paper from the author inscribed in the text.

These are first lines of Pasteur's *Mémoire*:

I feel I must point out in a few words how it came about that I undertook my study of fermentations. Having until now directed all my efforts toward attempting to discover the relations that exist among the chemical, optical, and crystallographic properties of certain substances, with the objective of shedding light on their molecular constitution, it may seem surprising that I should take up

a subject dealing with physiologic chemistry apparently quite remote
from my first labours: nevertheless, it is very directly related to
them. (Pasteur 1858: 404)

The person who wrote these lines is the real Pasteur. The "I" in the
first line, however, is not the real Pasteur (who is dead by now for
more than a century), but the *inscribed author*, the enunciator *in* the
text, a character in the play to be performed. He introduces himself
as someone with a background in studying the properties of certain
substances to shed light on their molecular constitution, that is, as
a chemist. When his text, read as a script, is turned into a play, we
may expect him to move around on stage in a laboratory, surrounded by Erlenmeyer flasks and other laboratory equipment.

Apart from introducing the inscribed author, the text also introduces an *inscribed reader*. At the point where Pasteur (the inscribed
author) states "it may seem surprising" that he took up a subject
quite remote from his first labours, the real reader may already be
bored, or distracted, but the text introduces a reader that will be
surprised. The *inscribed reader* is a figure controlled by the text.

In a play we expect some action. In semiotics, 'action' is a very
general term indeed: an action is any enunciation or performance
that has an effect; the term refers to any movement or pass (as in
football) that makes a difference to the state of a situation. The
semiotic definition of action is as minimal as possible. The term
does not imply that intentionality is involved in the action; neither
is the distinction between 'behaviour' and 'action' (a topic discussed at length in philosophy of social sciences and in philosophy
of mind) introduced. The semiotician's definition is not limited to
humans. Both humans and nonhumans can perform actions. When
a hammer hits a nail on its head, the hammer performs an action.
The generic term *actant* is introduced to cover any human actor or
nonhuman entity that acts, that is, performs an action. So a hammer
hitting a nail is an actant.

To analyse actions, we have to distinguish different *actantial
roles*, namely *object*, *operative subject* and *passive subject*. (As the
object represents some value – it is desirable or not to the subjects
– it is alternatively called 'value object'.) In football, when player
A passes the ball to player B, the ball is the object, player A the
operative subject and player B the passive subject. When a burglar
steals a document from a civil servant's home, the document is the
(value) object, the burglar the operative subject and the civil servant
the passive subject. The action, the burglary, makes a difference:

we move from a state in which the civil servant had the document and the burglar not, to a state in which the burglar has come in possession of the document and the civil servant has been deprived of it. Also the object has changed properties – from a document that was kept in tight security to one that the burglar may sell. A *translation* has taken place: the state of the operative subject, the passive subject and the object have changed, they have been redefined, that is, acquired a new meaning. If you want to follow a football match, keep your eyes on the ball and see how one player becomes the star of the evening, while other players are defeated. Likewise, to understand a play and to read a text as a script, focus on the *trajectory* of the *main circulating object*, that is, record the chain of *translations* that make up what is happening in the text and watch how operative and passive subjects interdefine each other.

Before Pasteur, other scientists had already studied lactic fermentation. In his *Mémoire*, (the inscribed) Pasteur dutifully refers to this previous work, to remark that the explanation of the phenomena remains unclear. He reports that it had been observed that in due course a deposit is formed of chalk and nitrogenous material with spots of a grey substance. But very often the grey substance is so mixed with the mass of casein and chalk that there is no reason to suspect its existence. Under a microscope, it is hardly possible to distinguish it and nothing indicates that it is a separate material or that it originated during the fermentation. Is organized life involved? "Minute researches have been unable to discover [it]. Observers who have identified some organisms have at the same time found that they were accidental and detrimental to the process" (Pasteur 1858: 408). So (the inscribed) Pasteur concludes: "the facts then seem very favourable to the ideas of Liebig [. . .]" (Pasteur 1858: 408). Nevertheless, he was led to an entirely different view.

To detect what led him to conclude that the chemists had it wrong, we have to follow him – that is: the Pasteur inscribed in the *Mémoire* – in his laboratory. He starts with producing "a good, ordinary lactic fermentation" (Pasteur 1858: 410) to extract a trace of the grey material that has formed in the deposit. He prepares a new liquid, a complex solution of albuminous and mineral material, and carefully filters it. Subsequently,

[a]bout fifty to one hundred grams of sugar [. . .] are [. . .] dissolved in each litre, some chalk is added, and a trace of the grey material [. . .] is sprinkled in [. . .]. On the very next day a lively and regular fermentation is manifest. The liquid, originally very limpid, becomes

turbid; little by little the chalk disappears, while at the same time a deposit is formed that grows continuously and progressively with the solution of the chalk. [. . .] After the chalk has disappeared, if the liquid is evaporated, an abundant crystallization of lactate of lime [calcium lactate] forms overnight, and the mother liquor contains variable quantities of the butyrate of this base [a by-product]. If the proportions of chalk and sugar are correct, the lactate crystallizes in a voluminous mass right in the liquid during the course of the operation. Sometimes the liquid becomes very viscous. In a word, we have under our eyes a clearly characterized lactic fermentation, with all the accidents and the usual complications of this phenomenon whose external manifestations are well known to chemists. (Pasteur 1858: 410; PH: 119)

In semiotic terms, in this passage, Pasteur (the inscribed author), is the operational subject; he is the one who adds sugar and chalk and sprinkles in a trace of the grey material. The passive subject is the liquid. It will turn from limpid to turbid. Where is the ball, the object, which with Pasteur will score his goal? It is the trace of grey material that is sprinkled in.

Having sprinkled in the trace of grey material, Pasteur finds "on the very next day" a lively and regular fermentation. To isolate the lactate of lime that has been formed in abundance, Pasteur evaporates the liquid. What is left, "[v]iewed as a mass, [. . .] looks exactly like ordinary pressed or dried yeast. It is slightly viscous, and grey in colour" (Pasteur 1858: 411).

The grey material Pasteur had sprinkled in has multiplied overnight. Subsequently, Pasteur isolates the grey material and finds – provided one doesn't dry or boil it in water – it is active. Very little of it is necessary to start *another* fermentation.

It can be collected and transported for great distances without losing its activity, which is weakened only when the material is dried or when it is boiled in water. Very little of this [grey material] is necessary to transform a considerable weight of sugar. (Pasteur 1858: 412; PH: 119–120)

What *is* this 'grey material'? It looks like ordinary yeast and like brewer's yeast one needs only very little of it to induce fermentation of sugars. After carefully comparing the two, (the inscribed) Pasteur concludes that it *is* a yeast, like brewer's yeast. However "one should not conclude from this that the chemical composition of the two yeasts is identical, any more than that the chemical composition of two plants is the same because they grew in the same soil" (Pasteur 1858: 412–3; PH: 120). "Here we find all the general characteristics of brewer's yeast, and these substances probably have organic structures that, in a natural classification, place them in neighbouring species or in two connected families" (Pasteur 1858: 412; PH: 120). So (the inscribed) Pasteur gives the ferment a name – 'lactic yeast' – and concludes that it is – like brewer's yeast – an organic (living) material.

What had started off as a nonentity, "often [. . .] so mixed with the mass of casein and chalk that there would be no reason to suspect its existence" (Pasteur 1858: 409), has been isolated, made clearly visible, has shown its power to initiate another fermentation and has been given a name and an identity (PH: 121). From a nobody, it has become a well-defined object, an actant that has revealed its organic nature.

Lactic fermentation, which used to be a slow, uncertain and obscure process, has become a process that can be mastered and accelerated by introducing the grey material in purified form.

> Chemists will be surprised at the rapidity and regularity of lactic fermentation under the conditions that I have specified, that is, *when the lactic ferment develops by itself*; it is often more rapid than the alcoholic fermentation of the same amount of material. Lactic fermentation as it is ordinarily carried out [by Pasteur's predecessors] takes much longer. This can easily be understood. The gluten, the casein, the fibrin, the membranes, the tissues that are used contain an enormous mass of useless material. More often than not these become a nutrient for the lactic ferment only after putrefaction [. . .] that has rendered the elements soluble and assimilable. (Pasteur 1858: 412; PH: 121)

By a chain of translations, the passive subject (the liquid) has changed from being limpid to turbid, the circulating object has changed (from an obscure grey mass to something that has a clear identity), the operative subject (Pasteur) has changed into someone who has made a discovery, and even the chemists (the inscribed readers) have changed (they will be surprised at the rapidity and regularity of the lactic fermentation under the conditions Pasteur (the inscribed author) has set up).

What have we been looking at? Commenting on the *Mémoire*, Latour identifies Pasteur's paper firstly as (the script of) an *ontological drama*, a play in which an at first sight irrelevant entity, an obscure grey mass, has entered the stage, to triumph, after having undergone several trials, by showing its power and true nature.

To what extent can we say that Pasteur has *discovered* the microbiological agent that causes lactic fermentation? Or should we say that by performing several operations he has *fabricated* it? To answer these questions, we turn our attention to the other drama Latour extracts from Pasteur's *Mémoire*, an *epistemological drama*.

Again, we need some terms from semiotics, viz. *shifting in* and *shifting out*, two important operations in narrating a story. Look again at the first lines of Pasteur's *Mémoire*, "I believe I must point out in a few words how it came about that I undertook my study of fermentations." This sentence asks the (inscribed) reader to shift his attention away from the 'I', the inscribed author, to his background, which in the lines that follow is immediately filled in as chemistry. We are invited to enter the workplace of a chemist.

A few paragraphs later, at the start of a new section called 'History', Pasteur writes: "Lactic acid was discovered by Scheele in 1780 in soured whey. His procedure for removing it from the whey is still the best one can follow" (Pasteur 1858: 406). This sentence asks us to move in time and space from Pasteur's laboratory to 1780 Sweden, where Scheele worked. We are shifting to a new frame of reference. (If we want to actually perform the play, we need a scene change.) Several other predecessors who have worked on fermentation are subsequently mentioned in the *Mémoire*. Having done with the history of fermentation research a few paragraphs later, we will again shift back to Pasteur's laboratory, to watch him perform the experiments discussed above. But something has happened in these moves. By now it is clear that Pasteur (the inscribed author) will deal with a phenomenon several other scientists have studied. By shifting out from his laboratory to the history of the field, and shifting in again to his own lab, Pasteur has added realism. Other people have discovered lactic acid and some of them have also seen the deposit of barely visible grey matter after lactic fermentation. *In* the text, an *internal referent* to the concepts of lactic acid and fermentation is provided. The inscribed reader is shown that a well-known phenomenon will be explored. If a real reader still has doubts, he may go back to Scheele's work.

At the end of his paper, Pasteur (the inscribed author) reveals that he started off his inquiries with the hypothesis that this grey matter is a living organism. But how can he *prove* that this is a fact? How does he *demonstrate* that the actions of a nonhuman agent, the yeast, rather then his own (human) interventions account for the outcome of his experiments? In other words: how does Pasteur deal with the problem that puzzled logical empiricism, that is, the decomposition of his conclusion into a conventional dimension and an empirical one?

Basically by setting up his experiments. Latour comments (PH: 129): "the experiment creates two planes: one in which Pasteur, the narrator, the inscribed author, is active, and a second in which the action is delegated to another actor, a nonhuman one." At first, Pasteur is active. As we have seen above, he prepares a complex solution of albuminous and mineral material, dissolves fifty to hundred grams of sugar, adds some chalk and sprinkles in a trace of the grey matter he has extracted from "a good, ordinary lactic fermentation". To make his claim to have *discovered* the existence of lactic yeast as an organism, Pasteur has to shift out from his plane to the ferment's plane. And so he does: he observes that the newly formed grey mass can be collected and transported and he sprinkles it again in a liquid he has prepared. He goes to sleep. Pasteur withdraws from the scene, to find the next morning that while he was sleeping the ferment has independently performed an action: an abundant crystallization of lactate of lime has been formed overnight. So both Pasteur and the yeast have been active: Pasteur has mixed and added materials and sprinkled a trace of the grey stuff; the yeast, having been well fed in the medium that Pasteur has provided, has grown to a clearly visible mass that allows Pasteur to collect and transport it, to start new fermentations and to speak about it in clear terms.

Latour comments: "what is at stake in the text is precisely the reversal of authorship and authority: *Pasteur authorizes the yeast to authorize him to speak in its name*" (PH 132). Yes, Pasteur is active, but so is the nonhuman ferment. Pasteur's activity allowed the nonhuman ferment to show its activity, while the nonhuman ferment subsequently allowed Pasteur to report that it is a real entity – one that can be manipulated, transported over distance, to still showing its activity.

Philosophers who are scientific realists claim that the entities described by correct theories do exist; electrons, fields of force and black holes are as real as toe-nails, turbines and volcanoes; their opponents, anti-realists and 'instrumentalists', claim the opposite: they are just fictions, tools for thinking and ordering phenomena. Hacking (1983: 22–23) provides a simple criterion for what made him a realist: "If you can spray [entities], then they are real." Of course, there can be discussions about *what* has been sprayed; but by then *that* it is real is no longer the issue. This is precisely what Pasteur did in his second experiment to back up his claim that lactic yeast exists.

A scientific paper weaves many statements into an argumentative narrative. Latour claims that a semiotic analysis of the narrative (i.e. by reading Pasteur's *Mémoire* as if it was the script of a play, analysing its 'scenography') reveals the 'ontological drama' and the 'epistemological drama' in which both the experimenter, the human actant, and nonhuman actants (in Pasteur's case: the lactic yeast) play roles. Both the human and the nonhuman actants have been active; they have interacted, redefined and translated each other. In the hands of the competent experimenter, fabrication and realism are not incompatible. The semiotic analysis of Pasteur's *Mémoire* allows us to understand why at the end, what is reported as a fact is both fabricated *and* real.

Having shown the activity of both Pasteur and the yeast allows Latour to divert from Collins' 'Sociology of Scientific Knowledge'. Scientific facts are constructed, but not only *socially* constructed.

> If we ignore Pasteur's [i.e. the experimenter's] work, we slip into [logical empiricist's] naive realism from which twenty-five years of science studies have tried to extract us. But what happens if we ignore the lactic acid's [i.e. the nonhuman's] delegated automatic autonomous activity? We fall back into the other pit, as bottomless as the first, of social constructivism, ignoring the role of nonhumans, on whom all of the people we study are focusing their attention, and for whom Pasteur spent months of labour designing his scenography. (PH: 132)

2.4   Realism in and about science

So, *what is science* and *what's so great about science*?

Scientists study reality, nature, to gain knowledge. That is what they say and that is what epistemologists suppose them to do. They want to determine the structure of chemicals released in improbably small volumes in the brain that stimulate or inhibit hormonal processes and behaviour; they want to know what causes lactic fermentation; they want to detect gravity waves emitted after a supernova explosion. However, in none of these cases the phenomena that they want to study are open for direct inspection and observation. They are too far away, barely visible, hidden from view, or not detectable with normal human senses. To seriously work on what interests them, scientists have to translate their interests into a problem, or a set of problems, that literally fits on a lab-bench, a desk, on a few pages or a computer screen. Well, even a desk may be too large; "desks encourage time-wasting activities," as Francis Crick used to say (cited in Sulston and Ferry 2002: 9).

An enormous amount of time and energy has to be spent on establishing the long chains that link the events or processes that interest scientists to the 'data' scientists work on. To become the object of observation, study, manipulation and discussion, the phenomena or processes in which scientists are interested have to be translated into figures, tables, graphs or other suitable forms. Instruments are used, liquids prepared to allow 'a trace of grey material' to grow, rats injected with synthetically produced complex compounds of peptides, assays set up, enormous antennas built. Inscriptions, 'immutable and combinable mobiles', translate matters into figures and texts, allowing data to be easily transferred to another place. They ensure that the chain is reversible and that if doubts arise it is possible to go back to the original material to do new measurements, calculations or tests. Once 'data' have arrived on a scientist's desk, they are available for being included in texts that may eventually be published in a journal, to be discussed and modified by other scientists. If doubts proliferate, scientists will lose interest and they will turn their attention to more promising suggestions. However, if a point is reached where all modalities have been dropped, a fact has been established. A stable connection between the remote entities and phenomena that the scientists want to study and the statements that they report about them will have been constructed.

This is science 'in the making': building long reversible chains that allow us to see further, deeper into what is unavailable to the normal human senses; chains that allow us to go back, and to know how to interact with remote entities, that is, "to act at a distance" (SA: 219 ff), to try to manipulate, change, or control what had been hidden to us before. This is what is great about science: it allows us to become familiar with – and often to master – things and events which are too far away, barely visible, hidden from view, or not detectable with normal human senses – endocrinological processes in human brains, events in distant star systems, the microbes that cause fermentation.

To build such chains is a human affair, taking time and effort. Nobody questions this. The question of what has been achieved once these chains have been set up, however, is subject of debate. How to interpret the epistemological status of scientific statements that result from all this work? Scientists and most philosophers of science claim that they correspond with 'brute facts' that *exist* in the world, independently of language and the work invested in scientific research – or in the modest version, they present the best correspondence available. Sociologists of scientific knowledge eschew the very idea of un-interpreted facts. As we have seen before, they claim that "[t]he apparent independent power of the world is granted by human beings in social negotiation" (Collins and Yearley 1992: 310). According to the sociologists of scientific knowledge, scientific knowledge does not represent 'brute' facts, but does state 'social constructions'.

Latour rejects *both* positions. He argues that we have to revise not only our ideas about scientific knowledge, but also about reality. To understand scientific practice, we have to study the *joint history*, the 'co-production', *of knowledge and reality*, to study *both* sides of the relation between subject and object, and not only the side of knowledge, understanding and social negotiation. And to do so, we have to focus on the 'circulating object', to trace the many translations that go into establishing the chains.

So from now on, Latour had to defend his claims on two fronts: not only against the charges of realist philosophers who conceived him as another postmodernist who considered scientific knowledge as just another socially constructed system of beliefs, but also against growing opposition from the sociologists of scientific knowledge.

It should not come as a surprise that sociologists of scientific knowledge were not amused when they discovered that Latour and Woolgar had eliminated in the second edition of *Laboratory Life* (LL2) the word 'social' from the subtitle of the book and that, in subsequent publications, Latour emphasized that to understand scientific practice, nonhuman actants have to be granted a role in the construction of scientific knowledge too.

The sociologists of scientific knowledge soon found the point where they diverged from Latour. In their view, Latour naively and mistakenly took inscriptions to be immutable, that is, to be not subject to 'interpretative flexibility'. To allow nonhumans to play the role Latour attributes to inscriptions means to ignore the role of interpretations and of interpretative flexibility, to accept naive realism and to conceive the authority of science to be based on nature, rather than on the social processes within the scientific community. Citing a case study about 'chemical transfer of learned behaviour' that found that "the inscriptions produced by mass spectrometers in biological laboratories were not universally accepted as representing reality," Collins and Yearley (1992: 311) argued that even on this low level interpretation comes into play and hence that human agreement has to be reached. To see inscriptions as immutable mobiles "is exactly how they are meant to look like to those who are less than expert. [. . .] To experts, everything is mutable" (Collins and Yearley 1992: 311).

However, Collins' and Yearley's argument is based on a elementary logical error. That *each* inscription is mutable, that is, that each link in a chain of translations can become the subject of controversy about its proper interpretation, doesn't imply that *all* of them are mutable. (Compare: that *each* lottery ticket can win the first prize does not imply that *all* lottery tickets can win the first prize). No doubt, the functioning and interpretation of an inscription device can be questioned at any time – within a laboratory, by a laboratory assistant, the scientist who is responsible for his work, or by one of his colleagues. But in many cases, scientists will take inscriptions produced by their instruments or handed in by their technicians as 'black boxed' data. They will open the 'black box' and question inscriptions only when they find a reason to do so – for example when irregular or completely unexpected results appear, or when malfunction of the inscription device or a human error is suspected. A scientist who starts to distrusts *all* of his laboratory equipment and procedures will soon become totally baffled. In no time his research project will have ground to a halt.

Interpretation *does* play a central role in scientific practice, however, but on another level, at a later stage. Once a paper is published in a scientific journal, the fate of its statements is in the hands of its readers. They may add modalities and give other interpretations of the results. Discussions about the adequate interpretation may go on for some time. Even when all modalities have been removed and agreement about interpretations has been reached, the results that the author claims as facts are – *epistemologically speaking* – still interpretations. However, at this time, Latour suggests, an inversion takes place: from now on, the research process will be narrated as the search for this particular outcome: what the paper states are the facts that scientists have been looking for, the facts that by now have been discovered to *exist*. The statements that express the paper's results are taken to *refer* to these facts.

In performing this inversion, scientists have used – what Kant called – their Reason. Limited to knowing the world only by their faculties of understanding and perception, the scientific community has reasoned that on the other side of appearance (Kant 1956 [1787]: xxvi–xxvii) there are 'things-in-themselves', that is, objects that *exist*. They may be wrong, of course. But for the moment, they have reasons to accept that the statements on which the scientific community agreed state facts that exist.

Kant paid mankind a great compliment: humans are bestowed not only with the faculty of understanding but also with Reason. But does an appeal to 'reason' convincingly account for the 'inversion' that has taken place? Not for an empirical philosopher. To *what* object does the scientific community refer?

If Pasteur (1858: 417) declares "that the new yeast is organized, that it is a living organism" and his colleagues have accepted this statement as a fact, what *is* 'the new yeast' he is referring to? Is it the barely visible grey mass in 'ordinary lactic fermentation'; the trace of grey material Pasteur sprinkled in the liquid he has prepared; the matter that he found the next morning, which can be collected and transported for great distances without losing its activity and that under the conditions that Pasteur has specified, allows fermentation to happen at a rapidity and regularity that will surprise chemists? Or perhaps 'lactic acid bacteria' (plural), that is, a clade of Gram-positive, low-guanine-cytosine content, acid-tolerant, generally non-sporulating, non-respiring, either rod- or cocci-shaped bacteria that share common metabolic and physiological characteristics – as the ferment is known today? Or all of them?

Having read Pasteur's paper in the right mode, we know what is referred to, namely the *circulating object* in the paper and if we want to include 'lactic acid bacteria' as well, also the circulating object in the papers of those who took up his work to develop it into modern microbiology. In the course of the experiments, it has changed properties – it has changed from being barely visible to being clearly visible – it has changed its name – from 'grey material' to 'lactic yeast' to 'a clade of bacteria' – and it has changed identity – from an unknown substance to a particular family of microbiological organisms. To understand what Pasteur's statement – that he has discovered 'a new yeast', a 'living organism' – *refers to*, we have to take the whole chain of translations into account in which this object circulates.

Certainly, not every chain of translations will deliver proof. An essential property is that the chain must be *reversible*. The succession of stages must be traceable, allowing for travel in both directions. "If the chain is interrupted at any point, it ceases to transport truth – ceases, that is, to produce, to construct, to trace, and to conduct it. The word 'reference' designates the quality of the chain in its entirety" (PH: 69).

Pasteur knew this. So he took the trouble to set up his chain of translations to meet this requirement. He made clear that he is dealing with a phenomenon that had already been investigated by Scheele and others. He too produced a good, ordinary lactic fermentation. And once he had isolated the abundant grey material that has grown after having sprinkled a trace of the grey material into his carefully prepared liquid, he showed that again it causes fermentation. Comparing it with brewer's yeast, he concluded that it is another form of yeast.

Has he achieved by this stage *adequatio rei et intellectus*, the holy grail of realists? No, Pasteur didn't discover a 'brute' fact. To observe the microbes' power and to identify the ferment as a yeast, a living organism, he had to set up experiments, to isolate, manipulate, and sow grey material, to *change* reality. He had to perform a lot of 'ontological work' first. Pasteur had to be active, before he was able to "authorize the yeast to authorize him to speak in its name." So are Collins and Yearley (1992: 310) right to claim that "the apparent power of the world is granted by human beings in social negotiation"? No, they are wrong on two scores. In the first place, scientific practice is not only a matter of agreeing on or negotiating interpretations. To limit our attention only to social, interpretative processes means neglecting all the 'ontological work' that is required. Secondly, what Collins and Yearley call 'the power of the world' is not 'apparent'. The activity of the lactic yeast, its multiplying in the culture that Pasteur had prepared, was necessary too. Without it, Pasteur would be left with nothing but traces of unknown grey matter. To conclude that fermentation is caused by a living organism, *both* Pasteur and the microbes had to become active to eventually become connected in a long chain of translations.

"Humanly speaking, let us define truth, while waiting for a better definition, as a statement of the facts as they are," Voltaire (2006 [1764]) wrote. Voltaire is right. For everyday human purposes Voltaire's definition will perfectly do: the statement 'snow is white' is true if and only if snow is white. But scientific statements are much more complicated than that. They have meaning only in theories, in the narrative of scientific papers and in the practical work of scientists that led to the results described in these papers. They pose not only the *epistemological* questions that have been widely discussed in philosophy of science and more recently in the sociology of scientific knowledge. To account for the practice of science, we have to extend our attention also to *ontological* questions. The idea that scientific statements refer to a reality consisting of 'brute facts' or 'things in themselves' that – like Lego bricks – build up to complex compounds may be tempting and for many sciences may also have served as a reductionist guiding principle. But it is too simplistic to account both empirically and philosophically for the *practice* of science, that is, the work done in laboratories and other scientific institutions and for the knowledge that is produced. To understand what science really is, what makes it so 'great', and to discuss its place in society, we need to focus also on the other side of the equation, to redescribe what reality is. We need a richer *ontology* than the doctrine that 'brute facts' lie waiting to be discovered by smart scientific minds.

To develop that ontology, would become Latour's major concern.

3

Science and Society

Social studies of science developed in the 1980s as an academic discipline against a backdrop of public concerns about science, technology and the role of expertise in democracy. To better understand the cognitive authority attributed to science and the interactions between science, technology and society, sociologists of science set out to study 'science in context' (Barnes and Edge 1982).

But what is the 'context' of science? For sociologists, the answer may be obvious: 'society'. However, this is far from undisputable. For centuries, it had been claimed that what characterizes science is precisely that it produces objective, de-contextualized knowledge, that is, knowledge the validity of which can be decided independently of when, where and by whom it was developed. Although after Kuhn this claim had to be nuanced, still it doesn't follow that 'society' is the proper 'context' in which to understand science. For Kuhn, to contextually understand science meant relating the achievements of contemporary scientists to the science of their predecessors and to social processes internal to science, not to 'society'.

Even if we grant sociologists that it might be interesting to study science in the context of 'society', what exactly are we talking about? When it came to listing societal conditions that might help explain the development and nature of scientific knowledge, in spite of their radical epistemological programme, social studies of science fell back on the standard repertoire of mainstream sociology to cite the usual suspects: interests, power, ideologies, and institutional structures. In *The Pasteurization of France* (1988), Latour argues that in taking such 'social facts' for granted, the sociologists of knowledge showed too little ambition. To understand the role of science in society, Latour claims, we need not only to rethink science, but also to redefine society and sociology.

The Pasteurization of France, an adapted translation of *Les Microbes* (1985), explicitly situates itself in the emerging discipline of sociology of scientific knowledge. The first part of the book is a historical-sociological study of Pasteur's role in late nineteenth and early twentieth-century France. But soon the study turns on sociology itself. "If sociology wishes to be the science of 'social facts', then it cannot understand this period" (PF: 40). Recapitulating his ideas in philosophical terms in the second part of the book, Latour provides an explicit ontology – a conceptualization of what the world and society *is*. To formulate, defend and reformulate this ontology and its consequences for social science would become a main concern for Latour for the years to come.

The Pasteurization of France takes as its starting point the fact that in France and elsewhere Pasteur is honoured as an undisputable genius who revealed the medical and economic potential of experimental biology, who changed agricultural, veterinarian, and bio-industrial practices, revolutionized public health, made cities habitable and by achieving all of this helped to substantially increase standards of living. How is it possible that one man has such immense influence? What made Pasteur great? Or do we completely misunderstand his role and that of his science when we attribute this power to him? These are the questions Latour wants to answer in the first part of *The Pasteurization of France*. The subtitle of that first part – *War and Peace of Microbes* – hints at what inspired his approach: Tolstoy's (2010 [1869]) great novel *War and Peace*.

Tolstoy's *War and Peace* narrates the story of how the individual lives of the members of five Russian aristocratic families became entangled with history when Napoleon invaded Russia in 1812. But *War and Peace* is not just a novel, and still less a historical chronicle. The novel deals with what in the words of C. Wright Mills (1999 [1967]: 31–32) the social sciences are all about, namely "[. . .] to help us understand biography and history, and the connection between the two in a variety of social structures." In an epilogue, struggling with the concept of 'free will', Tolstoy tried – not very successfully – to outline a programme that would do for history what the Copernican revolution had done for astronomy. In fact, what he was after was already implied in his novel. Tolstoy's style and his numerous reflections inserted in *War and Peace* provide substantial answers to the key-questions of the social sciences: 'What can one man do?', 'What is history?' and 'How to write history?'

How should we write history? With the advantage of hindsight, with a bird's eye view, a historian can start off by presenting the 'historical context' of the events about which he wants to write. He may also observe that while at a certain place actions took place, 'meanwhile', elsewhere, events happened that changed the context. But real-time actors don't have this advantage. Narrating the individual lives of his characters, if in time or space distant events or conditions play a role, a novelist has to introduce such events and conditions *into* the scenes of his story. Tolstoy does so, for example, by having a character recalling something from the past or by injecting his characters with commonplace ideas and clichés. And to signal that important events have occurred elsewhere, he introduces a messenger arriving at the scene to deliver the news about these events. But memory and clichés are not very reliable sources, whereas the news the messenger brings may be only a rumour, or may have been made up for showing off. Thus, in real-time, if 'context' plays a role, it is always bound up with uncertainties and mixed up with other concerns that require attention. So, in the opening scene of *War and Peace*, we find its main characters attending one of St Peterburg's famous *soirées*, being engaged in civilized conversations about politics, career moves and personal connections, the upcoming *fête* at the English ambassador's residence, rumours from France, the beauty of the ladies and the uniforms of men, all on a par. At this time (1805) in Russia, Napoleon is a subject for amusing stories and war a topic for civilized discussion; 'history' is still far away, at a distance.

What is history? For Tolstoy, the novelist, 'history' is neither a sequence of actions or events, nor social structures imposing their force on the lives of individuals. It is a series of *infringements* of the course of life; history appears when people are confronted with a novelty that affects their existence. Any such infringement will incite uncertainties. What has happened? Is it a rumour or a fact? What is going to happen? How will it affect us? Everybody suffers from lack of oversight; misunderstanding and misinformation are abundant on both personal levels and with regard to what is happening in the world at large. While a young lady may quietly wonder whether the man in the gorgeous uniform who has just entered the room will be her future fiancé, and an ambitious young man may try to persuade somebody to arrange for him a meeting with some important figure who might further his career, rumours from Paris are discussed and speculations about their significance burgeon.

Uncertainty reigns not only in these elite circles, convened at a *soirée*. Tolstoy shows that commanding army officers, including Napoleon and his Russian counterparts, are also fumbling in the dark. When the French and Russian troops got engaged in what was to become the battle near Borodino, both Napoleon and his Russian counterparts erred in locating the exact positions of the enemy. Of course, they had discussed strategy and designed grandiose plans for attacks. But their strategies were based on unreliable information and if they had been executed as planned, they probably would not have attained the intended effect. Surely, with pomp and gusto Napoleon played the role of 'Napoleon', the genius who is in charge and who knows exactly what he is doing, the hero of future history books. But Tolstoy shows that the orders generals sent out to their troops were frequently either misunderstood, or no longer appropriate to the situation – for example because the situation had completely changed as the enemy had unexpectedly decided to execute some actions of its own before the orders had arrived at the front. In fact, the orders may even not have arrived at their destination because the officer who had to deliver them got lost in the fog or was shot on his way. So, once the fighting had started, with tremendous loss of human lives, 'strategies' had to be adapted, redesigned, or discarded, again and again.

So, what can one man do? The truth is: very little on his own. Tolstoy shows that men act in uncertainty and that it is often unclear who is doing the action – what is horse and what is cart and whether the horse is before the cart, or the cart before the horse. Moreover, the narrative shows that a multitude of actors has to be taken into account. While senior officers were engaged in factional conflicts about strategy, the troops were not only fighting the enemy, but also had to deal with fog, snow, ice, diseases and the sheer vastness of Russia.

Tolstoy concludes that if we find in the accounts given us by historians successes explained by the ingenuity, power or virtue of leaders, and wars and battles by their conforming to previously prescribed plans, the only conclusion to be drawn is that these accounts are not true. 'Power', 'virtue' and 'genius' are useless as explanatory concepts. Instead, they are the *explananda*, the phenomena to be explained.

This sets the road Latour will follow in *The Pasteurization of France*.

> It takes Tolstoy some eight hundred pages to give back to the multitude the effectiveness that historians of his century placed in the virtue or genius of a few men. Tolstoy succeeded, and the whole of recent history supports his theories as to the relative importance of great men in relation to the overall movements that are represented or appropriated by a few eponymous figures. This is true at least where *politicians* are concerned. When we are dealing with *scientists*, we still admire the great genius and virtue of one man and too rarely suspect the importance of forces that made him great. We may admit that in the technological or scientific fields a multitude of people is necessary to *diffuse* the discoveries made and the machines invented. But surely not to create them! The great man is alone in his laboratory, alone with his concepts, and he revolutionizes the society around him by the power of his mind alone. Why is it so difficult to gain acceptance, in the case of the great men of science, for what is taken as self-evident in the case of great statesmen? (PF: 13–14)

In *The Pasteurization of France*, 'Pasteur' is dismantled in the same way Tolstoy has dismantled the 'Napoleon' of nineteenth-century historians. Latour presents the Pasteurian revolution not as the rational progression of the ideas of a genius triumphing over backward minds and reactionary forces, but as the outcome of innumerable actors who in the process defined each other, disease and society. The first part of *The Pasteurization of France* presents, analyses and reflects upon the shared history of microbes, microbiologists and society. The philosophy underlining this analysis is outlined in the second part of the book.

3.1   'The Pasteurization of France: War and Peace of Microbes'

The first, empirical part of *The Pasteurization of France* is based on a semiotic analysis of three periodicals, the *Revue Scientific* (from 1870–1919), *Annales de l'Institut Pasteur* (1887–1919), and *Concours Médical* (1885–1905). The corpus Latour has worked on is vast, but also limited. Foreign publications and – with a few exceptions – Pasteur's lab-notes and correspondence are left out. Moreover, semiotics is stripped to its bare bones: "since it is too meticulous to cover a period of fifty years and thousands of pages, the semiotic method is here limited to the interdefinition of actors and to the chain of translations" (PF: 11).

The authors whose publications are analysed have various backgrounds. A first group, the authors of *Revue Scientific*, consists of the 'hygienists' – the participants in the social reform movement that aimed to 'regenerate the nation' by improving national – especially urban – public health by a whole array of measures. Having no clear idea what causes illness, they enthusiastically set off in any direction: from improving sewage systems to therapies based on the cleansing and vivifying effects of water, country air and simple foods, to "methods of developing among the labouring classes a spirit of thrift and the saving habit." The congresses of the hygienist movement, Latour comments, "were like an attic in which everything was kept because sometime it might come in handy" (PF: 20). The second group, the 'Pasteurians', consists of Pasteur and his co-workers, writing in the *Annales de l'Institute Pasteur*. A third group, publishing in *Concours Médical*, a periodical of a medical association, consists of a variety of medical doctors – army doctors, general practitioners and other medical professionals. Each of these groups shows a style and interests of its own.

Who are the 'inscribed readers' of the articles in these periodicals? Basically, the colleagues of the authors. The articles in *Revue Scientific* address hygienists, while physicians publishing in the *Concours Médical* wrote for their medical colleagues. Time and again, the texts speak about 'we'. Pasteur and the Pasteurians, however, are the exception. The contributions in the *Annales d'Institut Pasteur* address hygienists, biologists and doctors alike, to show they have common interests.

The main problem the hygienists had to face was the 'variation of virulence', the mysterious patterns of the outbreak of diseases that proved contemporary doctrines of contagiousness to be inaccurate. For example, why are some fields – as farmers said – 'cursed', causing cattle to die of anthrax time and again, while other, neighbouring, fields were not? Existing contagious theories were of no help to explain this. Precisely on this point, Pasteur came to the rescue.

The hygienists started to refer to Pasteur and his discovery of micro-organisms in their articles already at an early stage. They believed him even before it had been proven that the measures Pasteur had shown in a few cases to be effective might be useful to society at large. For example, on the basis of Pasteur's first field experiments only, one hygienist exclaimed "anthrax will soon be a thing of the past." When Pasteur had cured only one child of rabies, another wrote "and now that we can cure rabies, we only have to expand and facilitate the treatment." "Yes, gentlemen, the day will come when, thanks to militant, scientific hygiene, diseases will disappear as certain antediluvian animal species have disappeared," another hygienist had already declared a year before (PF: 27–28). Latour comments:

> The confidence in the 'way laid down' by Pasteur must therefore derive from something other than the facts, hard facts. The confidence was not one that came only *from* Pasteur, but one that flowed back *on* Pasteur. [. . .] [Pasteur] had only to open his mouth, and others would turn his results into generalizations about every disease. [. . .] The social movement into which Pasteur inserted himself is a large part of the efficacy attributed to Pasteur's demonstrations. (PF: 28)

The medical professionals showed initial reluctance to embrace Pasteur's work. However, their hesitance disappeared when they too discovered Pasteur to be on their side. Army doctors first. Dealing with large groups of recruits rather than individual patients, and thus concerned with large-scale preventive measures, they soon seized upon Pasteurism with the same avidity as hygienists (PF: 114). Civilian doctors remained reluctant for longer. They took great care to separate what was exaggerated from what was useful from their own professional point of view: "Of all that slowly accumulating work, a body of precise knowledge will certainly emerge one day [. . .]. But we should maintain a certain reserve for the time being and not see bacteria everywhere, after previously seeing them nowhere" (PF: 117). Only in 1895, when the organization had been set up to bring the diphtheria serum to the physician's consulting room, did the physicians' resistance weaken. By then the *Concours Médical* advised medical doctors to change course:

> It may be not too soon to look ahead into the future that the scientific revolution, brought about by the beneficent discoveries of the illustrious Pasteur and his school, has in store for the medical profession. [. . .] [W]e shall take possession of the new ideas. [. . .] [W]e shall ask our young competitors [i.e. the Pasteurians], at the patient's bedside or in consultations, to share the benefits of their recent studies with us; at the same time let us tell them that, by way of compensation, we shall share with them the fruits of our experience in the skill of the medical profession. (PF: 129–131)

There is nothing new under the sun, sociologists may conclude. People follow their interests and they will assimilate new ideas only when they find them to be beneficial to their own purposes. However, sociologists will be surprised to hear Latour's far more radical conclusions.

Which messages did Pasteur and his co-workers bring to the communities of hygienists and medical doctors? That "we cannot form society with the social alone" (PF: 35); and that "the laboratory is an indisputable fulcrum" (PF: 72) to shift the balance of power between men and disease. These messages are the outcome of three steps.

In the first step,

> Pasteur or his disciplines visited in person distilleries, breweries, wine-making plants, silkworm rearing houses, farms, Alexandria decimated by cholera, and later, with the *Institut Pasteur*, all the [French] colonies. [. . .] They moved but remained men of the laboratory. They brought their own tools, microscopes, sterile utensils, and laboratory logbooks, using them in environments where their use was unknown. On the other hand, they redirected their laboratories to respond to the cause of those who they visited. (PF: 75–76)

So, to study anthrax, Pasteur went to the countryside, to collect on site materials at farms in the département Eure-et-Loir.

Back in Paris, knowing that predecessors had identified the anthrax bacillus but had been unable to show convincingly that it caused the disease, Pasteur performed the second step of his programme. To show that the bacillus was the sole agent of anthrax, he took just a drop of a culture liquid extracted on site from animals, to start a new culture, and to repeat this procedure several times. He then showed that the last drop of the last culture still caused the complete disease. So, he concluded, the bacillus is the unique agent of disease.

As these were still only laboratory experiments, in the third step Pasteur moved back from Paris to the field, to find out what happens in the countryside itself. How did the bacillus end up in an animal? Why were some fields 'cursed', while others were not? To answer this question, Pasteur concerned himself with techniques of burying the animals that had died of the disease. As the animals lost blood at the moment of burial, they also lost the bacilli. Koch had already shown in 1876 that they might live on as spores that retain their pathogenicity for a long time. But how did they turn up years later on the surface? "The Académie will be surprised to hear the explanation of this," Pasteur would report. "Earthworms are the messengers of the germ and from its deep burial place bring the terrible parasite back to the surface of the soil" (PF: 78). His conclusion led to a fairly obvious and simple prophylactic measure: animals that had died of anthrax must never be buried in fields intended for grazing or for growing fodder. If this measure was followed, Pasteur predicted, anthrax could be a thing of the past.

In three steps, using his laboratory as a fulcrum, Pasteur showed farmers the way to stop the dreaded cattle disease. But in the same move, he also solved the problem that had paralyzed the hygienists: the variance of disease. Pasteur's experiments showed the cause. Although dead animals were buried deep into the soil, earthworms brought the agent back to the surface, to infect new animals.

So Pasteur also had another message: we do not know who are the agents who make up our world, a message Latour sums up in one sentence: "there are more of us than we thought" (PF: 35). 'Social' relations do not form a society alone. "When we speak of men, societies, culture, and objects, there are everywhere crowds of other agents that act, pursue aims unknown to us, and use us to prosper. We may inspect pure water, milk, hands, curtains, sputum, the air we breathe, and see nothing suspect, but millions of other individuals are moving around that we do not see" (PF: 35). Only by taking a detour to the laboratory, can our unknown enemies be identified and become controlled. Once we learn from laboratory practices how to adapt agricultural, industrial and medical practices we may not only control the enemies inside the laboratory, but also in society at large. That message fell on fertile soil. The hygienists could begin to clean up their attic, medical doctors to reorganize their practice. Pasteur had shown them the way to pursue their own goals. However, we cannot 'explain' their actions merely by their political motives and interests, by 'social facts'. To understand how the Pasteurians revolutionized public health, veterinary practice and eventually society, we have to include the microbes in the narrative.

By moving the microbes to his laboratory, Pasteur succeeded in turning them from an unknown, invisible, dangerous enemy into a clearly identifiable one that could be defeated by vaccines or – e.g. by taking the role of earthworms into account – precautionary measures. Only by teaming up with Pasteur's laboratory, which enabled them to take the existence of the microbes into account and to redefine their concerns and actions, could the hygienists and the medical doctors redefine their old interests. Turning themselves into a new kind of professional, they became engaged in political games renewed "from top to bottom with new forces" (PF: 40). The detour to Pasteur's laboratory turned the scales. Before Pasteur, farmers and veterinarians were weaker than the invisible anthrax bacilli. The disease was unpredictable, reinforcing the idea that idiosyncratic local circumstances were its cause. In Pasteur's laboratory, the bacillus was identified as the unique cause of the disease. Returning to the field, pointing to the role of burying practices and earthworms, Pasteur had shown how to account for the variation of the disease. In three steps, Pasteur showed how man could become stronger than the bacilli. But the scales were also turned in another way. By embracing and implementing Pasteur's ideas and techniques in their practice, the hygienists also redefined Pasteur to become the great man of French science who revolutionized veterinary practice, public health and society.

Taking stock at the end of *War and Peace of Microbes*, Latour claims to have given back to the sciences "the crowd of heterogeneous allies which make up their troops and of which they are merely the much-decorated high command whose function is always uncertain" (PF: 147). The allies include hygienists, army doctors, laboratory equipment, farms and earthworms. They "were an integral part of so-called scientific objects. [. . .] We understand nothing of the solidity of a fact if we do not take into account the unskilled troops" (PF: 147). In Tolstoyian fashion, Latour thus takes issue with the myth that scientific knowledge flows from the mind of a genius to spread through society under the flag of reason.

> Microbes play in my account a more personal role than in so-called scientific histories and a more central role than in so-called social histories. Indeed, as soon as we stop reducing the sciences to a few authorities that stand in place of them, what reappears is not only the crowds of human beings, as in Tolstoy, but also the 'nonhuman', eternally banished from the [i.e. Kant's] Critique. (PF: 149–150).

Latour has a message not only for historians and philosophers, but also for sociologists eager to explain scientific developments by power, ideologies or other social facts.

> If sociology wishes to be the science of 'social facts', then it cannot understand this period [of the Pasteurian revolution]. If [. . .] we still

call ourselves sociologists, we must redefine this science, not as the
science of the social, but as the science of *associations*. We cannot say
of these associations whether they are human or natural, made up
of microbes or surplus value, but only that they are *strong* or *weak*.
(PF: 40)

Irreductions, the second part of *The Pasteurization of France*, attempts
to expand and generalize these lessons in philosophical terms. But
before he turns to them, in a reflexive note, Latour realizes that he
cannot claim for himself what he denies to the Pasteurians. He has
spoken of the Pasteurians as they spoke of their microbes. "My
proofs are no more irrefutable then theirs, and no less disputable.
I must go looking for friends and allies, interest them, draw their
attention to what I have written [. . .]" (PF: 148). The reader is
warned.

3.2 'The Pasteurization of France: Irreductions'

Irreductions, Latour's *Tractatus Scientifico-Politicus* (PF: 234), is an
abstruse text. It consists of apodictic, often flamboyant, decimally
numbered statements and aphorisms, interspersed with interludes
of personal reflection. But not only its style makes *Irreductions*
sometimes hard to penetrate. Latour also struggles with the problem
encountered by anyone who tries to communicate a radical idea:
to express and explain new ideas, one has to use old words. One
may try to use them in new combinations, use scare quotes and
new allusions to stretch their meaning and add a few neologisms,
but by diverging too far from the established usage of language
one will soon call down upon oneself the suspicion of writing
gibberish.

In Latour's case, the problem is aggravated for two reasons. In
the first place, many of the key-terms in *Irreductions* – like 'force',
'strength', 'potency' and 'reality' – have a wide variety of meanings
not only in everyday language but also in science, politics and the
philosophical tradition. For example, where 'force' is used in combination with 'interest', the term suggests an agent with intentions
and a political reading of the term; however, in combination with
'lever' and 'fulcrum', a physical reading is suggested, without any
allusion to intentionality. In *The Pasteurization of France*, Latour uses
'force' in both combinations, leaving the reader puzzled about
what is meant. In the second place, Latour's ontology is strikingly

different from the intuitions and the conceptual bifurcations that
in the long tradition of Western philosophical thought we have
been taught to hold dear. Reading that nonhuman agents have
wishes and interests, one may easily think that the author is out of
his mind. Not only old habits, but also philosophical notions die
hard.

The introduction to *Irreductions* explicitly addresses the latter
point. Again a novel serves as a guide. Tournier's (1972 [1967])
Vendredi ou les limbes du Pacifique inverts Defoe's story of Robinson
Crusoe. Where Tolstoy's *War and Peace* deconstructed the hero of
nationalistic historiographies, Tournier takes issue with the Euro-
centric and racist *Weltanschauung* of Crusoe, to replace it with the
worldview of *Vendredi* (Friday).

> Crusoe thinks he knows the origin of order: the Bible, timekeeping,
> discipline, land registers, and account books. But Friday is less
> certain about what is strong and what is ordered. Crusoe thinks he
> can distinguish between force and reason. As the only being on his
> island, he weeps from loneliness, while Friday finds himself among
> rivals, allies, traitors, friends, confidants, a whole mass of brothers
> and chums, of whom only one carries the name of man. Crusoe
> senses only one type of force, whereas Friday has many more up on
> his sleeve. (PF: 154)

When Friday has carelessly blown up the powder house, Robinson
Crusoe finds his neatly ordered world destroyed. In Tournier's
version of Defoe's story, instead of rebuilding his stockades and
turning back to his orderly discipline, Crusoe decides to follow
Friday to discover that the latter lives on a completely different
island. Crusoe too has to learn that he is among a wide variety of
friends, allies, and rivals. Soon he feels at home in this world. When
more than 28 years after the shipwreck that brought Crusoe to the
island a new boat, the *Whitebird*, has anchored off their island,
Friday departs. Confronted with the brutality, hatred and greed of
Whitebird's captain and his crew, and realizing the irrevocable relativity of the goals they feverishly pursue, Crusoe decides to remain
behind. To his surprise, the next morning he finds the young Esto-
nian ship boy on the island, who has fled the ship. He decides to
call him *Jeudi*, "the day of Jupiter, the god of Heaven, and of Children's Sunday [the fête of laughter and play]" (Tournier 1972
[1967]: 253).

Latour didn't have to wait for a Friday to blow up the ordered
world of modernity. He had done so himself in the first part of *The*

Pasteurization of France. Subsequently, in *Irreductions*, like Tournier's Crusoe, Latour decides not to restore the old order, but to describe a world of a multitude of forces engaged in trials, that may enlist each other to gain strength and that signify their reality by resisting trials. In this world, man is only one of many forces. Like Tournier's Crusoe, Latour is satisfied with his new world. He decides to be "as agnostic and as fair as it is possible" (PF: 236), wondering in the last line of *The Pasteurization of France*, whether, like Tournier's Crusoe, he too will find new company. "In the old Europe are we still capable of swapping places in this way?"

Force, strength, trials – the language suggests a Hobbesian world of wars of all against all, a Machiavellian arena of actors driven by Nietzschean Wills to Power. Latour is aware that the bellicose associations may scare his readers. He suggests that replacing 'force' by 'weakness' may ease them (PF: 252, n.6). In place of 'force' we may talk also of 'entelechies', 'monads', or more simply 'actants' (IRR: 1.1.7).

Ignore the allusions to Aristotle and Leibniz; Nietzsche, who Latour had read as a student, comes closer. His philosophy resounds in the epiphany Latour had on his way from Dijon to Gray in 1972 (PF: 162–3; cf. ch. 1). In the 1880s, in notes that were posthumously published under the title *Der Wille zur Macht*, "Nietzsche [. . .] claimed that nothing in the world has intrinsic features of its own and that each thing is constituted solely through its interrelations with, and differences from, everything else" (Nehamas 1985: 82). "[Kant's] 'thing-in-itself' is nonsensical. If I remove all the relationships, all the 'properties', all the 'activities' of a thing, the thing does not remain over," Nietzsche (1966: 563) wrote. "The properties of a thing are effects on other 'things': if one removes from thought all other 'things', a thing has no properties, that is, there is no thing without other things, that is, there is no 'thing-in-itself' " (Nietzsche 1966: 502–503). Like Nietzsche, Latour gives *relations* pride of place over essences. This explains why he can suggest that instead of 'strength' his readers may read 'weakness' – both are properties that only make sense in a relational context: something has 'strength' or 'weakness' only in comparison to something else – and why one may read 'actant' instead: an entity is an 'actant' if and only if its action makes a difference to a situation and to other actants. So "[n]othing is, by itself, either reducible or irreducible to anything else" (IRR: 1.1.1). What an entity *is* depends on its relations with other entities, the web of connections in which it has become established, the translations that it has performed and the translations

that have been performed on it. But each attempt to act or to engage
in a relation can fail. To try, a verb, is Latour's point of departure
(IRR: 1.1.2). Each relation involves a trial. To exist, to have properties, to show activity, an entity has to subsist, to hold its ground in
tests, to resist in trials; existence is not guaranteed by an essence.

To get a feel for the ontology Latour introduces in *Irreductions*,
conceive a play staged in a theatre as a model for the world. What
is happening and who and what do exist on stage? To answer
these questions, one has to focus on the relations between actors;
they define what each is and how they change in the course of the
play, to become a hero, a villain or the abandoned lover. How
does a main actor become a character in the world on stage? Only
by interacting with other actors and with artefacts, and by speaking about himself, events, other characters, and artefacts. What
kind of character is he? In a play, on stage, artefacts, other characters and his own actions will define him. The words he speaks,
the crown on his head and the fact that other players address him
as 'your majesty' define the actor as a king. To become a character
in the play, the actor hired by the theatre-company needs props
and other actors. The web of relations he will engage in on stage
constitutes what – in the world enacted on stage – he *is*. In a subsequent stage performance, with other props, other players and
other lines being spoken, the actor will become another character,
another being.

However, if the actor that the theatre company has hired moves
clumsily, and instead of making what was intended as a kingly
gesture causes the crown to fall in the orchestra pit, the world
created on stage loses its king. Being not good enough to do his
act, the actor lacks the necessary strength to be the king on stage.
Likewise, when other members of the cast forget their lines, or
when the cardboard crown turns out to be much too big to fit the
actor's head, the king on stage loses his reality. All that remains is
a failed performance. So, to become part of the world, to exist and
stay real, one has to get enough strength to 'resist' trials, to hold
one's ground in a test. Chances of success may be improved by
enlisting another actant to become a third actant, or by organizing
several actants in a chain or network (IRR: 1.4.2). But enlisting or
enrolling others will always come with costs: any alliance will
translate an actant into a new one.

With Nietzsche, Latour defends a *relationist* ontology. But how
to analyse these relations empirically? Semiotics taught Latour the
vocabulary and the technique for extracting what *is* (in a text read

as a play) from narrative forms. The ontology Latour introduces in *Irreductions* is semiotics writ large.

No intrinsic features, but a multitude of human and nonhuman actants; no a priori distinction between humans and nonhumans, man and nature. Everything that Latour grants to humans – force, action, strategies, interests – he also grants to nonhumans. An uncommon world indeed.

How many forces or actants are there? We will have to find out by trials, that is, by measuring actants against each other. The actant we call Pasteur has to meet the actants he calls microbes, to set up a trial, an experiment, to allow them to multiply in his carefully prepared cultures, to turn them from an invisible enemy into a visible one that he can manipulate – isolate, dry, transport – at his will. So both the subject and the object of an enquiry are actants and we had better abandon the terms 'subject' and 'object' because they will set us on the course of epistemology. *That* something is real is defined by its resistance in a trial. *What* it is, is not given by an eternally given essence, but also shows in trials. To know what a fish is, biologists, the fishing industry, and consumers, all have to set up trials. And for each of them, a fish is something different. While for the one a fish *is* a vertebrate, for the other it *is* a commodity, or *is* food. This is not a matter of different 'interpretations'. To know that the food on your plate at dinner is fish, you don't need to 'interpret'; just start eating and enjoy it. If you still remain suspicious about what you are eating, ask the chef or call for somebody to do a few chemical or biological tests. You can leave your textbooks on hermeneutics untouched. There is no need to add an extra layer of 'interpretations' apart from reality.

Latour's philosophy conceives knowledge-production as a move, as (a set of) translations that have to be realized *in* the world, rather than as an attempt of a subject, a human mind, or a community of scientists, to acquire knowledge *about* 'real' objects, that is, to interpret phenomena. "Nothing is known – only realized," *Irreductions* (1.1.5.4) states rather cryptically (cf. also PF: 148). However, the phrase is less cryptic than it might seem at first sight. Do you want to know whether something is strong? Try to bend it. Do you want to know whether A is longer than B? Set up a trial: compare the two. Is what's on your plate really fish? Perform, or order, a test. Knowledge is a matter of actants setting up trials in a world populated by other actants to make what is real – that is, resists trials – visible.

"There can be no 'symbolic' to add to 'the real'" (IRR: 2.5.6.2). "It is not possible to distinguish for long between those actants that are going to play the role of 'words' and those that play the role of 'things'" (IRR: 2.4.5). Latour is "willing to talk about 'logic', but only if it is seen as a branch of public works or civil engineering" (IRR: 1.4.4). There is no contrast between force and reason. "What we call 'science' is made up of a large array of elements whose power we prefer to attribute to a few" (IRR: 4.1.6). That is what the first part of *The Pasteurization of France*, *Laboratory Life* and Latour's analysis of Pasteur's *Mémoire* tried to show. Analysing the sequence of texts from three periodicals, in *The Pasteurization of France*, Latour asked *ontological* questions: what does the world we encounter in these texts contain? Which actants do we encounter and how do they translate and interdefine each other in the course of fifty years? How was 'Pasteur' constructed, how did he come into being as the famous son of France?

But is it possible to *empirically* reconstruct by semiotic methods what has happened in France around 1900 from three periodicals and even to claim to have detected the role of the "crowd of heterogeneous allies" of Pasteur, including microbes and earthworms? Semiotics brackets off both external referents – the things and events a text speaks about – and the pragmatic or social context of authors and readers (cf. ch. 2). It instructs us to analyse how *in* and *by* a text internal referents, inscribed authors and inscribed readers are constructed. So can a semiotic analysis ever bring us to the historical past? We'll get to that in § 3.3.

Latour describes the world, knowledge and society in terms strikingly different from the Western tradition that runs from Plato to Descartes and Kant up to modern philosophy of science and the sociological tradition. In the second half of *Irreductions* he starts explicitly to distance himself from these traditions, again in apodictic statements and aphorisms. He will do so extensively, in more detail and more clearly, in *We Have Never Been Modern*. We will leave these issues for discussion in chapter 5. For now, let's first see what has been achieved, and at what costs.

3.3　Another turn after the social turn

The Pasteurization of France got mixed reviews. The first half of the book and its style were widely praised. According to the philosopher of science Hacking (1992: 511), Latour had produced "an imaginative and well-informed account of what Pasteur did, although the form of the story is certainly nonstandard. It is rich in quotations that other scholars ignored." Vernon (1990: 345), a historian of science, concluded that the first part was "engagingly written, though there is little new in substance. Many historians already knew that, in the short term, little in the practice of medicine, hygiene or brewing was changed by Pasteur or germ theory; we would expect that hygienists and Pasteurians would use each other for their own ends. But these points are worth emphasizing and Latour does it very well, not least by exposing the presumptions implicit in 'heroic' accounts." Reviewing the French version, *Les Microbes*, for *Social Studies of Science*, Knorr-Cetina (1985: 585) wrote: "Latour has the rare gift of stimulating one's thinking and of giving ideas, even if one does not agree with what he says! But the greatest interest of the book, and its wider relevance to social studies of science, lies perhaps in the fact that this is the first attempt by a student of scientific practice to begin to analyse the question of the social 'diffusion' of knowledge, of the 'scientization' of society, or of 'knowledge utilization'."

However, the second half of the book, *Irreductions*, was met at best with puzzlement. Vernon (1990: 345) suggested that it "deserves detailed attention from philosophers and sociologists." Philosopher Hacking (1992: 511), however, limited himself to remarking that this half of the book "is a fascinating, wandering series of meditations and reflections which will certainly delight some readers, but which will be impenetrable to most readers of this [philosophy of science] journal." Knorr-Cetina provided more substance. She criticized at length "Latour's Nietzschean theory of the political nature of all social life [. . .]" (1985: 581). Moreover she commented that "[t]hose committed to the sociology of knowledge perspective must be disappointed by the fact that [. . .] the microbes appear in the book as a weapon in the political struggles between Pasteur and the others [. . .]. [They] are treated as natural agents and not [as] (social) constructions" (Knorr-Cetina 1985: 584).

The last comment set off an alarm. In a footnote to the English edition, Latour stressed that Knorr-Cetina and other reviewers of *Les Microbes* had misunderstood what he was up to. He explicitly denied that he was "interested in offering a social and political interpretation of Pasteur as an alternative to other cognitive or technical interpretations" (PF: 252 n.10). What he had intended was not another study in social studies of science, but to go further, to take the discipline to its next step, to make another turn after the social turn. "I am interested [. . .] in retracing our steps back to the moment when the very distinction between *content* and *context* had not yet been made. If I use the words 'force', 'power', 'strategy', 'interests,' their use has to be *equally distributed* between Pasteur and those human and nonhuman actors who give him his strength" (PF 252 n.10 italics added). Apart from the four methodological principles introduced by Bloor (cf. ch. 2), Latour insisted on accepting a fifth rule: in the analysis humans and nonhumans should be treated symmetrically. After the 'social turn' that the new sociology of scientific knowledge had introduced, another turn was called for.

It may have been a small step for Latour, but it was a big step for the social studies of science community – too big in fact. Not only did the new turn Latour suggested imply that the ambition to sociologically explain the content of science by its (social) context had to be discarded. The suggestion that human and nonhuman resources should be considered symmetrically also meant that the cornerstone of the sociology of scientific knowledge – the flexibility of human interpretations of the nonhuman world – lost its prominent analytical position. Latour seemed to suggest that in the construction of scientific knowledge nonhumans had independent power, a voice of their own. As a consequence, the turn Latour suggested was perceived as a reversion to a rather naive form of realism and as a retraction from the terrain that the sociology of scientific knowledge claimed to have conquered. Soon polemical, in some cases quite aggressive, reactions started to appear, arguing that what Latour had suggested as a step forward, in fact was a step backwards, a bad, reactionary joke (Amsterdamska 1990; Schaffer 1991; Collins and Yearley 1992; Bloor 1999a; 1999b).

To some extent, Latour had called this reaction upon himself. In the first place, by presenting his ideas as an extension of the programme of social studies of science, rather than as an explicit transition to another programme, which in fact it is. *Les Microbes* had already many references to the social studies of science literature; its English edition refers also to work in the sociology of scientific knowledge published after 1984, the year *Les Microbes* was published. At the same time, Latour distanced himself from the sociologists of scientific knowledge, for example by teasingly claiming that old-school (mostly American) sociology of *scientists* was more reasonable than the new (mostly British) sociology of scientific *knowledge* (PF: 257 n.27). The science studies people he considered as his friends did not appreciate the irony.

Also Latour's language offered little help to convince his friends. That microbes have 'interests', 'wishes' and 'strategies' and that they 'enthusiastically' grew in the medium that Pasteur provided, is hard to swallow for most readers. Living in the modern, rationalized world, they are used to making a categorical distinction between on the one hand the domain of humans and on the other hand the 'disenchanted' nonhuman world (Weber 1968 [1919]: 594). Instead of thinking that microbes 'enthusiastically' grew in the medium Pasteur provided, they think that Pasteur had just observed the microbes to multiply. Period. Referring to the microbes' 'enthusiasm' to do so was conceived as a misleading, if not bizarre, anthropomorphic addition. Perhaps French readers may appreciate peppering a sentence this way, but serious British academics do not. Latour's comment (PF: 260 n.5) that whenever he used the words 'interest' and 'interested' he was not referring to sociological 'interest theory', but to the notion of translation – that is, "simply what is placed 'in between' some actor and its achievements" – didn't help to convince his sociological colleagues either. "[I]f we check this claim by trying to substitute the words 'in between', or the general idea of in-betweenness, in the passages in which Latour uses the word 'interest' we find it does not work," observed Bloor (1999: 100), who found it unnecessary to try more charitable ways of checking Latour's suggestion.

Finally, it was unclear about what Latour was aiming at in methodological terms. He declared his intention as "to explain the science of the Pasteurians" (PF 8–9). But what does 'to explain' – a term that pops up in *The Pasteurization of France* more than a hundred times – mean? Not to analyse "the 'influences' exerted 'on' Pasteur or [. . .] the 'social conditions' that 'accelerated' or 'slowed down' his successes" (PF: 8). Instead, Latour declares his intention to set out to "describe [the science of the Pasteurians] without resorting to any terms of the tribe" (PF 9). However, in the chapters that follow, Latour cites at length from the texts of the Pasteurians and those who applied Pasteur's ideas and techniques, to rephrase only (meta-) terms like 'proof', 'efficacy', 'demonstration', 'reality' and 'revolution' that the tribes – and later historians – used to evaluate what Pasteur had achieved, in terms of 'force', 'strength' and 'trials'. Does that count as 'providing an explanation'?

"In order to make my case, I seem to be putting myself in an indefensible position," Latour sighed (PF: 9). Quite so, the sociologists of scientific knowledge thought. Latour's "criticisms are based on a systematic misrepresentation of the positions [of sociologists of scientific knowledge] he rejects, and [. . .] his own approach, in so far as it is different, is unworkable," Bloor (1999: 82) concluded. The new turn Latour suggested "is obscurantism raised to the level of a general methodological principle" (Bloor 1999: 97).

However, it is one thing to express irritation about what is perceived at first sight as abstruse language and obscurantist claims, it is quite another to prove that someone who suggests a new approach has made mistakes or has missed crucial points that affect his conclusions. The Cambridge historian of science Schaffer was one of only a very few commentators who took the trouble to confront Latour on his own terrain, to go into the details of his study and to point to aspects that had been left out of Latour's empirical study but that Schaffer claimed to be significant. Schaffer's (1991) extensive review of *The Pasteurization of France* deserves close reading for its detailed criticism of the semiotic method and of Latour's "attribution of purpose, will and life to inanimate matter, and of human interests to the nonhuman" – an inclination Schaffer called 'hylozoism', a polemical term he borrowed from the English poet and philosopher Coleridge (Schaffer 1991: 182).

Schaffer starts out by generously praising Latour for having succeeded "brilliantly" in demystifying Pasteur and for having adopted two "admirable rules of method" that the "Anglophonic reader, if moved by an irenic spirit" can also find in the work of Collins and other sociologists of scientific knowledge: "first study systems in the course of controversy, when all is unstable and up for grabs, since closure effaces the memory of the work through which the taken-for-granted is established; second, do not accept the rigid boundary between the scientific-technical and the social-contextual which is often a result of these passages of action, and so cannot be used to explain them" (Schaffer 1991: 177, 180). However, the main body of his review criticizes both Latour's use of semiotics that led him to neglect the social processes involved in the production of scientific knowledge, and Latour's 'hylozoism'. Schaffer claims that the two points are directly connected. "Hylozoism directs our attention away from the forces [the sociology of scientific knowledge has brought to the fore] which help close [a] dispute. It therefore disables understanding" (Schaffer 1991: 189). Both points of critique and their connection deserve scrutiny.

First Latour's use of the semiotic method. In the first part of *The Pasteurization of France* Latour analyses what the *inscribed readers* of three periodical are supposed to take as the message of the published texts. Unfortunately, instead of 'inscribed reader', Latour used the term 'ideal reader'. The phrase puts Schaffer on the wrong foot. He takes the ideal reader to be "a figure produced by *authors* as a role for their audience, whose behaviour the *author* must seek to control" (Schaffer 1991: 178 italics added), instead of a figure produced by the *text*. A passage in which Latour reports that "when I began to read the *Revue Scientifique* after the defeat of 1870 [of France in the Franco-German war], I was surprised to observe that little was said about Pasteur," adds to the confusion. "Are we to identify the 'ideal reader' with the nineteenth-century public, or with the author of *The Pasteurization of France?*" Schaffer (1991: 178) asks.

The answer is: with neither. Schaffer confuses the real readers of a text with the inscribed readers whom a semiotic analysis brings to the fore. When an article in the *Revue Scientifique* contains the phrase "Yes, gentlemen, the day will come when, thanks to scientific hygiene, disease will disappear", there is no doubt who are the 'gentlemen' the text addresses as *inscribed reader* (namely fellow hygienists). Of course, a *real* reader may still have had doubts about the power of scientific hygiene; he may not even have finished reading the sentence because he was called for dinner. Likewise, when in the *Concours Médical* a text states that "we should maintain a certain reserve for the time being", it is clear who the inscribed "we" are (namely fellow medical professionals) although, of course, there may have been real nineteenth-century subscribers to the *Concours Médical* who by that time already viewed Pasteur's work with less reluctance.

But does a semiotic reading that focuses on inscribed readers then allow one to conclude anything about what *real* nineteenth-century readers thought? Yes, by following the course of debates in *Revue Scientifique* and *Concours Médical* over decades, observing the *changes* in the published texts over time, Latour can conclude that *real* nineteenth-century readers indeed came to accept the messages that the *inscribed readers* of the articles were supposed to. Thus he is allowed to conclude that as a historical, empirical fact that from the early 1880s the hygienists trusted Pasteur without question and that – in due course – the medical professionals would do so too.

Schaffer has one more string to his bow. "I should remind the reader again at this point that I am limiting my sources to what an 'ideal' [i.e. inscribed] reader would know of Pasteur and his alliances, were he or she to read only the *Revue Scientifique*," Latour warned his Anglophone readers (PF 256 n.19). That leaves open the question how Pasteur knew who his 'alliances' were. The answer, of course, is that his experiments in his laboratory and his onsite observations (e.g. of burial practices), led Pasteur to conclude that microbes cause the diseases he studied. Schaffer correctly points out that to describe what Pasteur did in the seclusion of his laboratory, Latour had to fall back on Pasteur's lab-journals, that is, to follow a route that the readers of the *Revue Scientifique* and *Concours Médical* cannot have travelled. If we allow Latour to use these texts, is it clear who were Pasteur's 'alliances'? Pasteur may have believed that, but, as Schaffer is quick to point out, his work was subject to controversy. In the scientific community, Pasteur met strong opponents like Koch, Pouchet and others. The controversies in the scientific world are omitted from Latour's story; Schaffer (1991: 186, 188) even claims "deliberately" so.

Schaffer claims that this omission allows Latour to shortcut the *social* processes of overcoming (what Collins has called) the experimenters' regress, that is, the social processes that bring a controversy to closure and that turn local, private lab work into a public, established scientific fact. Pasteur's statement that the disease he has studied is caused by microbes became an established 'scientific fact' only after the controversies in the scientific community had reached closure, a process that is left undiscussed in *The Pasteurization of France*. This leads to Latour's other sin, Schaffer claims. "By suppressing the controversies which surround Pasteurism [and the *social* processes that bring them to closure], Latour is able to use 'the microbes' as wilful actors. Instead of symmetry [in dealing with true and false beliefs], he tries hylozoism. [. . .] Only through hylozoism can he speak of the events within Pasteur's lab's walls" (Schaffer *op. cit.*: 186).

So, according to Schaffer, Latour's 'hylozoism' is a direct consequence of his use of the semiotic technique of analysis. The semiotic method leads to ignoring the underdetermination of scientific proof (i.e. the flexibility of (human) interpretation) and the crucial role and social processes to reach closure in scientific debates. A true sociologist of scientific knowledge would have explored the ways in which the statements that came out of Pasteur's local laboratory are made into robust scientific facts, that is, would have described how in complex *social* processes the experimenters' regress was overcome. He would have provided an even-handed treatment of Pasteur and Koch, and of the interests of the French republican and German imperial regimes that matched their respective research strategies, Schaffer (1991: 191) suggests. In this account, the microbes would have no independent role. The explanation of the course history took would be framed exclusively in terms of human and national interests.

Would Latour's account have to be corrected and rewritten if the scientific controversies in which Pasteur got involved were included? Schaffer thinks so, but – within the confines of a book-review – he has to limit himself to making a few programmatic suggestions. Apart from dealing with the way the debates between Koch, Pasteur, Pouchet and others reached closure, he suggests a sociology of science and society would also have to "account for the ways the world must change to allow facts to travel" (Schaffer 1991: 191).

Indeed. But, of course, this is exactly what Latour claims to have done. He has studied how the facts Pasteur claimed to have discovered were accepted by hygienists, even before the scientific community had closed the debate about his claims. Pasteur's conclusion that there are more of us than we thought, that we have to take microbes and earthworms into account, did travel in hygienists' circles as an established fact, not as a disputable interpretation of his local lab-work that waited for approval by other scientists. As we have seen before, the hygienists believed Pasteur even before the measures he suggested had shown their efficacy in more than a few cases. Their confidence in him was not based on hard facts, established after the closure of controversies that divided the international community of early microbiologists. Pasteur did not have to wait for his French scientific opponents and for Koch to approve his work. His message that there are more of us than we thought was accepted by French hygienists and began to affect society even before the scientific controversy had reached closure. And, as noted before, this had an important effect: "the social movement into which Pasteur inserted himself is a large part of the efficacy attributed to Pasteur's demonstrations" (PF: 28). Hence, to describe the role of Pasteurianism in French society, Latour has good *empirical* reasons to ignore – "deliberately" or not – the controversies around Pasteur's work and the social processes in the scientific community that brought the controversies about his claims in scientific circles to closure.

Schaffer's review once more highlights the philosophical divide that was growing between Latour and the sociologists of scientific knowledge. By suggesting 'another turn after the social' Latour did not extend the programme of sociology of scientific knowledge; rather he moved to a different programme.

Sociologists of scientific knowledge studied the social processes of the production of scientific knowledge, that is, the production of statements *about* the world that after closure of disputes about the interpretation of phenomena are accepted as stating scientific facts. Their problematic was an *epistemological* one; they tried to explain in sociological terms how scientific knowledge was established. Although allowing also interests that are usually conceived as 'external' to science to play out in the social processes that turn local lab-work into publicly accepted facts, they nevertheless gave the scientific community pride of place. For the sociologists of scientific knowledge, a 'scientific fact' is established if and only if the scientific community has established closure of its controversies. But this leaves an important question unanswered. Who are part of the relevant community? Pasteur and Koch, for sure. But what about the hygienists, army doctors and general practitioners? They did more than just pave the way to allow facts to travel that had been fully confirmed in academic circles. They contributed significantly to Pasteur's success, even before closure had been reached in scientific circles. To understand the role of Pasteur *in* French society, a focus exclusively on knowledge production in the scientific (academic) community leads to short-sightedness.

3.4   The turn to ontology

Although he explicitly presented *The Pasteurization of France* as a contribution to the sociology of scientific knowledge, in fact Latour had set out on a different journey. He displays Pasteur's work not for its producing knowledge *about* the world, but as moves, that is, as (a set of) translations, that have to be realized *in* the world. Latour raises *ontological* questions: how are these translations realized and how do they affect the way a wide variety of actants interdefine each other? The tools that help him to answer these questions are suggested by semiotics. And to use them, we do not only have to "abandon knowledge about knowledge" (SA: 7); we also have to abandon what we think makes up the world. We have to be "as agnostic and as fair as possible" (PF: 236). We not only have to apply the four methodological principles the sociology of scientific knowledge adopted, to be impartial and symmetric with regard to truth and falsehood, rationality and irrationality, that is, to abstain from the evaluatory terms the people we study use, but also to abstain from *a priori* attributing different categories and competences to humans and nonhumans.

No doubt, interpretations play a role in the making of science and in the way scientists' work changes society. Both in *Laboratory Life* and in his analysis of Pasteur's *Mémoire*, epistemological questions and the role of interpretations are addressed. But for Latour, interpretations are not statements *about* the world, they are moves *in* the world; they are one type of many possible operations that actants use to redefine and enrol another. Of course, we may expect to find differences between human and nonhuman actants. Humans show many capacities that are not available to nonhumans. For example, they can interpret a text, an event, and phenomena; they not only have the capability to see something, but may also *see* something *as* something else (cf. Wittgenstein (PU, Part 2, XI), thus redefining what they encounter, exposing 'interpretative flexibility'. As their performances show, nonhumans display other capabilities, many of them not available to humans. *Bacillus anthracis* can kill cows and sheep, while *comma bacillus* can cause cholera and kill people. Inanimate objects have again other capacities. There is a multitude of actants in the world and there is a multitude of operations that they use to translate and interdefine each other. Interpreting is only one of them. Perhaps it is an operation that belongs exclusively to the domain of human actants – something that ethologists and biologists may question. But even if we have to *conclude* that interpretation is an exclusive human competence, this does not prevent us from approaching humans and nonhumans *methodologically* on a par, that is, symmetrically, when we *start out* to describe the world.

The history of science shows abundant evidence that getting rid of established and intuitively convincing distinctions may open up fruitful new ways of understanding. In the sixteenth and seventeenth centuries, astronomy boldly removed the Ancient distinction between what goes on in the heavens and in the sub-lunar world. Darwin's theory of evolution audaciously placed man in the animal kingdom. If counter-intuitive ideas had been forbidden in science, little if any progress would have been made. So why not be bold and try a methodology that treats humans and nonhumans

on a par, that is, one that does not impose *a priori* some asymmetry
between on the one hand human intentional action, interpretation
and reasoning, and on the other hand a material world of causal
relations, to see where this will lead us? If there are real differences
among human and nonhuman actants, we'll come across them;
they will show up in the course of the investigation, by the way
actants resist trials.

By suggesting 'another turn after the social' Latour proposes
to reject the conventional distinction between humans and nonhumans as the *starting point* for redescribing the world and
society; however, a distinction between full-fledged human
subjects and respectable objects may be the point of *arrival*
(PH: 182). But epistemologically it is a big step. It implies that
knowledge production is no longer conceived as a matter of for-
warding (exclusively human) interpretations of the world, but is
analysed as one of many moves *in* a world that consists of both
humans and nonhumans. By taking the turn Latour suggests,
ontology rather than epistemology gets pride of place. After this
turn, epistemological issues will be analysed in ontological terms,
that is, in terms of chains of translation that involve both human
and nonhuman actants. Then, as Latour's anatomy of Pasteur's
Mémoire shows, we find that in science, knowledge and reality are
'co-produced'.

The turn to ontology that Latour suggests redefines science
studies. It can no longer pretend to *explain* science in terms of social
processes and interests. To describe society and the role of science
in society, the analysis cannot be restricted to the social processes
of knowledge production. Social order, society, scientific communities, are made up of much more than only the 'social relations'
sociology-textbooks focus on. As Latour summed up Pasteur's
message: "There are more of us than we thought." One also has to
include the role of microbes and inanimate objects in the description. To describe science and its role in society one has to adopt a
different style of work and a new explanatory ideal.

To spell out this new ideal, Latour (1988b) contrasts two types
of explanation:

> The first is common to all disciplines: hold the elements of A [the
> *explanans*] and deduce – correlate, produce, predict, reorganize,
> comment or enlighten – as many elements of B [the *explanandum*] as
> possible. The second [is]: *display the work* of extracting elements from
> B, the *work* of bringing it to A, the *work* of making up explanations

inside A, the *work* of acting back on B from A. [. . .] The first is reductionist, because holding a single element of A is tantamount to
holding all the elements of B. The second [is] irreductionist because
it adds the work of reduction to the rest, instead of subtracting the
rest once the reduction is achieved. (Latour 1988b: 162–3)

The Pasteurization of France follows the irreductionist programme.
Latour has trailed the route Pasteur travelled to find the cause of
anthrax and the variation of the disease: the steps from the field to
his laboratory where experiments were conducted and back to the
field, showing in each step the *work* that was implied, the *translations* that were enacted and the *trials* that had to be overcome.
Displaying Pasteur's work and showing how it is taken up by
hygienists and medical professionals, Latour speaks "about the
Pasteurians as they have spoken about microbes" (PF: 148). He
describes what they did, what they encountered and how they
were changed along the way.
  Irreductionism is evidently a risky project. To engage in it,
"[a]ll the literary resources that can be mustered to render an
account lively, interesting, perceptive, suggestive and so on have
to be present" (Latour 1988b: 170). It requires a *style* that meets a
simple criterion for success: does the account *add* anything to the
world I thought I knew? Does it give a *richer* account than the
stories I have heard before? To paraphrase Foucault (1994: 3.540–1),
does it "make evident what is so close, so immediate, so intimately
linked to us, that because of that we did not perceive it?"
  Laboratory Life was written to a large extent in the style of an
ethnographic case study. Empirically oriented sociologists were
familiar with this style and for that reason they may be excused for
failing to see that Latour's aims and the semiotic methods that
underpinned the analysis differed radically from both traditional
sociology of science and the sociology of scientific knowledge. In
The Pasteurization of France, Latour has shifted his attention from
the production of scientific knowledge to the role of science and
technology *in* society, to enable its readers to learn how through
chains of translations a wide variety of actors, human as well as
nonhuman ones, become defined and redefined. To do so, he had
to follow the actants and to keep the evaluative repertoire of out-
siders – sociologists and historians – at arm's length.
  In *Aramis or the Love of Technology*, Latour's style became
even more explicit than in *The Pasteurization of France*. The book
is cleverly and elegantly staged as a *Bildungsroman*. A young

engineering student who wanted to spend a year at the *École des Mines* in sociology is introduced. He is asked to study *Aramis*, a French high-tech automatic train system – a smart combination of private cars and public transport in which cars that carry a small number of passengers are coupled to form a train, to be uncoupled at the right spot to bring the passengers to their preferred destination. After a very promising start, up to the point where the French Prime Minister enjoyed riding in one of its prototypes, *Aramis* was buried without fanfare. Who had killed *Aramis*? In the course of his attempts to answer this question, the student learns what technology and society are made of. How does he learn this? By collecting and analysing documents with technical details and evaluation reports, by conducting interviews with engineers and policy-makers, and by listening to the reflexive comments of his professor. Out of these fragments, the student reconstructs the world of and around *Aramis*. Like an archaeologist, from scattered splinters of the past, he constructs how content ('technology') and context ('society') co-develop. He comes to understand that *Aramis* is not a thing 'out there' that was killed by social forces (Capitalism, the Communist Party, Bureaucracy, or Economic Interests) that had been waiting in the dark to kill off *Aramis*. He comes to know that a properly functioning, stable public transport system requires for its subsistence a large network of both human and nonhuman actants that translate each other. And he discovers why this complex, fragile, hybrid being eventually disintegrated: it was not loved enough.

How do we, as readers of Latour's book, learn all of this? In the same way: by constructing a plot out of the fragments in the book. The literary form of the book supports its content. The book follows the student on his trips, provides the documents he has collected, excerpts of his interviews, the reflections of his professor. The various sources are distinguished by their typography. By selecting a clever, and very engaging format of writing, Latour tackles the problems usually framed as 'problems of method', to show the world of technology, our modern world, in a way we had failed to see before.

Science and technology studies showed the heuristic power of Latour's ontology, his vocabulary and ethnographic approach. Although he would continue to conduct ethnographies of science and technology (SA; PH; CB), in the 1990s, Latour began to explore the consequences of the successes of science and technology studies in wider domains. He set out to question the underlying assumptions of social science and of the mainstream of Western philosophy, and to address what he conceived as the main problem the world faces: the global ecological crisis.

Arguing that to understand the world we live in, social science (RAS) and our self-image as people who have a modern, rationalized worldview have to be revised (NBM), and that the values that we claim to hold dear need to be redescribed (AIME), Latour gradually turned into a public intellectual.

4

Another Social Science

Sociology, conceived as the science of 'social facts', cannot understand the period of the Pasteurian revolution, Latour argued in *The
Pasteurization of France*. Even the new, ambitious sociology of scientific knowledge he found to fall short. Claiming to explain the
content of science in terms of social processes, professional and
ideological interests, it completely missed the crucial role of nonhumans. This leads Latour to boldly state in *Reassembling the Social*
that "[s]ocial theory has failed on science *so radically* that it's safe
to postulate that it has *always failed* elsewhere as well" (RAS: 94).
To understand the modern world the whole conceptual apparatus
of social science needs reformulation.

Latour's harsh verdict on sociology is sugared only a bit.
He concedes that some of the intuitions of social science are
useful to build upon (RAS: 21–22) and that the statistical information the social sciences produce have made it possible for us
all to 'have' a society to live in (RAS: 226). Latour also admits
that in most situations it would be silly not to use the sociological commonplaces that populate sociology textbooks and that
one finds in journalism. Notions like 'Maori culture', 'totalitarian-
ism', 'lower-middle class' or 'social capital' offer convenient
shorthand to designate what is already assembled together. "But
in situations where innovations proliferate, where group boundaries are uncertain, when the range of entities to be taken into
account fluctuates" (RAS: 11), Latour claims sociology goes
astray. That is: in all except the most orderly, confined, static and
dull cases.

Latour's posture in regard to sociology is strikingly different from his attitude towards other scientific practices. While he has followed a wide variety of scientists and engineers to learn as an anthropologist from their practices, and has approached lawyers, ecologists and an ethnopsychiatrist in similar vein, when it comes to sociologists, he draws a firm line. He does not follow their activities, but advises them to radically change course and to revise their conceptual apparatus and methodology. According to Latour, sociology fails because it mixes up two entirely different meanings of 'the social': on the one hand it uses the term to designate the process of assembling a collective; on the other hand it uses the term to designate the outcome of these processes, which are conceived as ingredients of a specific realm, distinct from the realms of nature and individual minds.

Sociology emerged as an academic discipline out of attempts to reflect on the significant changes in Western societies the nineteenth century had brought. Should industrialization, urbanization, alienation, the rise of professions and of mass-movements be conceived as signs of the emergence of a new type of social order, as a fundamental shift from *Gemeinschaft* to *Gesellschaft*, that raises weighty moral and political questions?

In the nineteenth century, regional and state bureaucracies had started to generate "an avalanche of printed numbers" (Hacking 1990: 3). Their statistics made regularities apparent in the occurrence of criminality, suicide, alcoholism and other social phenomena. Apart from social life that had been known from earlier times – face-to-face encounters, families, village life, power struggles – the newly detected regularities were taken to be signs of the existence of a realm that hitherto had been hidden from view. To become aware of its features and to explain them, specialized methods, provided by statistics and the theories of a new discipline, sociology, would be necessary. Sociologists had found their vocation. In their desire to establish a proper discipline of their own, they claimed to have identified a domain of study of their own, namely 'society', 'the social', a realm that could be distinguished from the domains that concern natural scientists, economists, or psychologists and that would require its own methods to be studied.

How to study the newly detected social realm? The founding fathers of modern sociology, Weber and Durkheim, were both influenced by (neo-) Kantian ideas, the predominant philosophy of their time. Kant's influence on Weber shows in the way he formulates the object and methods of sociology. He defined the object of sociology to be 'subjective meaningful social action'. According to Weber, to study social action sociologists have to understand (*verstehen*) the subjective meaning an individual or a group attaches to an action. In contrast to Weber, mimicking what he conceived as the methods of the natural sciences, Durkheim embraced a positivist methodology; Kant's influence shows in his conception of the domain of sociology. What Durkheim singled out as the proper object of sociology, 'social facts', he called 'collective representations'. They are the sociological version of Kant's pure forms of intuition and categories: not the framework of a supposedly universal human mind, but sets of classifications, concepts, rules and rituals that define the way a group conceives itself in its relation with the objects which affect it (Durkheim 1968 [1901]: xvii).

Weber and Durkheim developed sociologies of radically different style. Their diverging approaches presented their successors with a host of methodological and conceptual puzzles: how to integrate Weber's bottom-up view on social order with Durkheim's top-down approach (cf. e.g. Berger and Luckmann 1967; Knorr-Cetina and Cicourel 1981; Giddens 1984)? But in spite of their differences, Weber and Durkheim agreed on one important point. They both conceived sociology to study a unique domain, distinct from the domains studied by other disciplines. Requiring 'subjective meaning' to be involved in social action, Weber distinguished sharply between social and natural events: "A collision of two cyclists [. . .] is something that merely happens, like a natural event. On the other hand, their attempt to avoid hitting each other, or whatever insults, blows, or friendly discussion might follow the collision, would constitute 'social action'" (Weber 1972b [1922]: 11). Conceiving society to be a realm *sui generis* (Durkheim 1968 [1901]: xvi), Durkheim argued at length that social facts can neither be explained in psychological terms, nor in terms of natural circumstances like climate and geography. In Weber's sociology, human subjects study other human subjects; in Durkheim's social science, social facts are explained by other – underlying, hidden – social facts. For both, mundane objects, technologies and other nonhumans were conceived neither as *explanandum*, nor as candidates for the *explanans*. The subsequent sociological tradition has followed suit.

Of course, no academic discipline can seriously claim to study the world *in toto*. The domain of enquiry has to be limited; matters that are not touched may be left to other disciplines. Division of labour is useful also in the world of science. Physicists don't have to study the role of catalysts in speeding up chemical reactions; chemists will take on that task (borrowing insights from physics to understand their role). However, an explicit omission of issues becomes questionable if it hinders the answering of the key questions that a discipline set out to address.

This, Latour claims, is the case with sociology abandoning the role of nonhumans and artefacts from their concerns. Sociologists set out to understand social order; but they are

> constantly looking, somewhat desperately, for social links sturdy enough to tie all of us together or for moral laws that would be inflexible enough to make us behave properly. When adding up social ties, all does not balance. Soft humans and weak moralities are all sociologists can get. The society they try to recompose with bodies and norms crumbles. Something is missing, something that should be strongly social and highly moral. Where can they find it? Everywhere, but they too often refuse to see it [. . .]. (Latour 1992: 227)

What do sociologists fail to see? Suppose you leave your home to have a private talk with a friend. Sociologists will point out that your action is 'subjectively meaningful' and that your action will be constrained by the 'collective representations' embodied in society's norms, laws and social structures. Is that all? No, of course it is not. It leaves out the obvious. Namely that your car will not start if you have not fastened your seat belt (a requirement of the law wired-into your fancy car); that you will limit your speed not only for moral reasons or for fear of the law, but because you know that the concrete slab of a 'sleeping policeman' will damage the suspension of your car if you drive too fast; that to have a private talk you will have to meet your friend in a secluded space, which your friend's home will provide; and that the architect has provided a clever solution to enter the walled space of your friend's home: you don't have to make a hole in the wall yourself, but you can enter through the front door; however, finding the door being closed, you will either need to push the button of the electric bell first, to wait for your friend to open the door, or to have a key to your friend's house.

In short, the obvious that sociologists fail to notice is that being a social, civilized, moral, law-abiding citizen who wants to have a private talk with a friend depends on much more than 'subjective meanings' and 'collective representations' – namely on a huge array of mundane artefacts and technologies. They constitute the "missing masses" that would help sociologists to solve the riddle of why – in spite of the weakness of the human will, our unsteady norms and morals and the liberties that the law allows – social order is remarkably robust (Latour 1992).

"Much like sex in the Victorian period, objects are nowhere said and everywhere to be felt. They exist, naturally, but they are never given a thought, a social thought" (RAS: 73). To include them in their studies, to investigate the way in which social order is the result of a wide range of assemblies of human and nonhuman entities, sociologists need to break away from the epistemological and ontological assumptions that laid the foundations of their discipline. They have to get rid of the idea that 'the social' designates a realm separate from the world of nonhumans (e.g. artefacts) and to abandon the idea that 'social action' is exclusively human, subjectively meaningful behaviour. Yes, having a private talk with a friend is a 'social' action; and the laws and norms you comply with constitute 'social' facts. But neither the action, nor these facts, are made exclusively out of 'social' ingredients. Those who think otherwise are making a category mistake, similar (to use Ryle's (1970 [1949]: 17–18) classic example) to the mistake of the foreigner who after having been shown a number of colleges, libraries, playing fields, museums, scientific departments and administrative offices in Oxford or Cambridge, asks 'But where is the University?' They think that the whole is necessarily of the same category as its parts.

'The social' is not a separate realm, Latour claims. It is a name for the *movements* to progressively compose a collective out of heterogeneous, human as well as nonhuman elements. So, instead of studying 'society' – the assembly of already gathered entities that sociologists believe have been made out of social stuff – we would do better to begin studying how – both human and nonhuman – entities are gathered together to form collectives.

Latour calls traditional sociology, the science that conceives society as a well-identified, separate realm and that provides understanding of 'social action' and explanations of 'social facts' in sociological terms, the 'sociology of the social'. His alternative, a 'science of associations', engaged in reassembling the collective process that makes up the social by tracing a wide array of associations and by analysing how stabilization is achieved, is called 'Actor-network theory' (ANT for short).

'ANT' had already become a common name to designate the empirical work in science and technology studies of Latour, Callon,

Law and others (Callon and Latour 1981; Callon 1986; Law 1994). At one time, Latour had written that on second thoughts the name was misleading and should be dropped (Latour 1999b). Alternative names, however, never caught on; 'ANT' was to stay to become the name of a new, flourishing approach in social science. Apart from ethnography, new digital research methods have been developed to help the exploring and visualizing of the controversies that emerge in the process of assembling (in particular socio-technical) collectives (Venturi 2009; Venturi 2010; Venturi et al. 2014; Venturi, Jensen and Latour 2015). Science and techology studies still provide most examples of the approach, but by now ANT has proved it merits also in other branches of the social sciences, as well as in geography, spatial sciences, architecture, and in business and organization studies.

The work of Callon, MacKenzie and others in ANT-inspired sociology of economics, in particular the study of markets, stands out (Callon 1998a; MacKenzie 2006; MacKenzie, Muniesa and Siu 2007; MacKenzie 2009; Çalışkan and Callon 2009; 2010). Studying in detail the role of 'market-devices' and economic models in the constitution of markets, they argued that economics, in the broad sense of the term, doesn't study given domains, that is, 'the Economy' or 'the market', but rather "performs, shapes and formats" (Callon 1998b: 2) them. The title of MacKenzie's (2006) pioneering study of models used in finance all over the world summarizes their role: they are "an engine, not a camera"; they shape and even constitute financial markets. A substantial part of today's financial markets, namely the multi-trillion dollar derivatives markets, would simply not *exist*, if finance theory and modelling had not been developed in the 1970s – a fact that got lost in most discussions about the 2008 financial crisis.

4.1   Deploying what makes up the social

Reassembling the Social, Latour's introduction to actor-network theory, interweaves criticism on the foundations and methods of the 'sociology of the social', lessons from science studies, conceptual exercises, and advice on how to do ANT-ethnography. The text is long, dense, somewhat repetitively and heavily footnoted. Latour knows he has to wrestle with the intuitions of his readers and the conceptual framework they have acquired in social science education. In an amusing interlude – an imagined dialogue between an

ANT-professor and an ambitious student from the London School
of Economics who has heard that by now ANT is the hottest thing
in town – the confrontation is made explicitly. But in contrast with
Socratic dialogues, the professor doesn't succeed in enlightening
the student. The student leaves the professor's office, to turn his
back on ANT. By then, the reader of *Reassembling the Social* is only
halfway through the book . . .

Actor-network theory is not an alternative theory *in* sociology
– that is, a theory competing with e.g. Giddens' (1984) 'theory of
structuration' or with Rational Choice Theory. ANT is an alternative *to* the 'sociology of the social', an alternative social science, a
technique for redescribing the social world by tracing the associations of humans and nonhumans that make up a 'collective'. Given
ANT's provenance in science studies, a further misunderstanding
is imminent. The most visible – and controversial – aspect of
Latour's work in science studies is the role he attributes to 'nonhumans'. Extending the lessons from this subfield to the whole of
social science might suggest that when sociologists study other
domains than science – politics, religion, the art-world, health care
or whatever other subject might interest them – they just have to
learn to take also some 'material' objects into account. No doubt,
they have to. But 'nonhumans' is not just an odd name for what
used to be referred to as nature, or the material world. The term is
introduced to allow agency for *anything* nonhuman. It may also
refer to entities of an entirely different nature than the ones one
encounters in Latour's analyses of science and technology. For
example, some religious people insist that they are being moved
by divinities, spirits and voices (RAS: 235); these are 'nonhumans',
no doubt, but quite different ones from – say – microbes and door-
bells. Likewise, in art, when actors explain at length that they are
attached to, moved, or affected by works of art that 'make them'
feel things (RAS: 236) they are certainly not referring to the chemistry of the paint, or the wooden frame of the picture. Actor-network theory is a technique for detecting how connections between
heterogeneous, human and nonhuman, entities make up a state of
affairs that we used to call 'social'. It is a technique for making
descriptions, not a shopping-list for reminding you that you have
to include also a few 'material' objects in your accounts.

ANT is a theory about *how to learn* what makes up collective life
– by letting the actors have some room to express themselves and
by being attentive to their enunciations. In engaging ANT, one has
to start as undecided as possible about what makes up the social.

Forget everything you thought you knew about collective life; be undecided as to what makes up the social; be prepared for surprises. "*Recording* not filtering out, *describing* not disciplining, these are the Law and the Prophets" (RAS: 55).

So what are we going to study? Face-to-face interactions, small groups, French culture, Dutch society, parliamentary democracy, McDonald's fast-food chain, the Roman Catholic Church, or globalization? What will be the level of our analysis: micro- or macrosocial phenomena, or perhaps the way these levels relate? At the outset, we may want to define the *scale* of the phenomena we are interested in.

First and foremost, we have to suppress this urge. The task of defining and ordering the social should be left to the actors themselves, rather than be taken up by the analyst. An organization, an institution, a group is not a given entity; a group is "the provisional product of a constant uproar made by millions of contradictory voices about what is a group and who pertains to what" (RAS: 31) and this also holds for whatever we usually take to be social aggregates. Spokespersons will 'speak for' a group's existence, trying to define its identity; they will have theories of actions both of themselves and of others; group boundaries will be negotiated with other ('anti-') groups. This is what we have to focus on. Scale is what actors achieve by scaling, spacing and contextualizing each other. Social aggregates should not be defined by ostensive definitions introduced by the analyst, but need to be defined by the *performative* definitions that actors provide. Their performances, that is, the *work* they have to invest in constituting a group, will have to tell us the scale of the group, that is, whether it is small or large. Social aggregates have ontologies of different scales, 'variable ontologies'. When Pasteur showed that there are more of us than we thought, he redefined the scale of society. To account for social aggregates, we need to start being uncertain about what a group, a society, an institution is and to focus on the controversies that arise with respect to its formation. By tracing the activities of *group formation*, the analysis will tell us what the appropriate frame of reference and scale are.

The same holds for 'actors'. In an ANT-account, as in semiotics, agencies are presented as anything that *does* something and that makes "some difference to a state of affairs, transforming some As into Bs through trials with Cs" (RAS: 52–53). Again, like in semiotics, the term *actant* is used for anything that modifies a state of affairs and makes a difference in the course of some other agent's actions. In an ANT account, what is doing the action is always provided "with some flesh and features that make them have some form or shape, no matter how vague" (RAS: 53). What form or shape? We'll have to find out. So an actant can be anything: Bush Jr, the US Army, the United States, imperialism, a drone, or the couch on which someone sits watching the horrible news from Afghanistan; and as we have seen before, also microbes, a doorbell or a key. Actants can be concrete or abstract, artificial, structural, anything – as long as they can be shown to make a difference. They can be humans or nonhumans (a term, we should remember, which does not designate a domain of reality but is introduced to allow agency for anything nonhuman). If you believe 'social structures' are doing something, then show *how* they make a difference. If they don't leave a trace, they should be eliminated from your account. "A good ANT account is a narrative or a description in which all the [actants] *do something* and don't just sit there" (RAS: 128).

The 'sociology of the social' limits itself to describing and explaining human social actions and the social structures that constrain their action. ANT "follows the actors in their weaving through things they have added to social skills so as to render more durable the constantly shifting interactions" (RAS: 68). To be more precise, two technical terms need to be introduced. An *intermediary* is what transports meaning or force without transformation; *mediators*, on the other hand, transform, translate, distort, and modify the meaning or the elements they are supposed to carry (RAS: 39). An intermediary, however complex internally, can be taken as a black box; defining its input is enough to define its output. In contrast, the input of mediators is never a good predictor of their output; their specificity has to be taken into account every time. The constant uncertainty over the intimate nature of entities – are they behaving as intermediaries or as mediators? – should be a major concern for the ANT-analyst.

In many cases, the nonhumans we commonly identify as 'material objects' and 'technical artefacts' shift from being mediators to intermediaries. They are doing nothing surprising; their behaviour is predictable and uncontroversial. The sociologists of the social may be excused – for a minute – to have taken their role for granted and to consider them as 'black boxes'. Of course, we take for granted that our friend's house has a front door. So why should one bother to mention that one has to enter his house through the door? But watch what happens when the doorbell is broken, or when you have forgotten to bring the key. Then surely, you will become aware of the door's role. Suddenly you discover that the door is *doing* something: it is obstructing your entrance to the house.

Precisely because of their role as intermediaries, nonhuman objects help to stabilize social order: their behaviour is predictable, actors can count on them; they present no surprises. To show their role as mediators, "to make them talk", the ANT-analyst may have to apply a few tricks (RAS: 80–82).

A first trick is to study innovation, the practice in which mediators are introduced or invented and in which they have not been stabilized, that is, not yet become a 'black box'. (It is no accident that science and technology revealed the role of nonhumans by studying laboratory life and scientific controversies.) A second one is to take distance. Archaeologists have to reinvent the use of the artefacts they have found on site; ethnologists have to do the same with the artefacts of other cultures; you may have found yourself in a similar situation figuring out the use of complicated electronic equipment without a proper manual. A third trick is to study accidents, breakdowns and strikes. To know nature, Bacon (quoted in Kuhn 1977: 44) advised to "twist the lion's tail" – the lion will surely show what he's up to. Accidents and breakdowns similarly provide a view on the mediating role of artefacts that under normal conditions will perform an intermediary role. Fourthly, one can bring the role of objects as mediators to light by using archives, documents and historical accounts. Finally, one can use one's imagination or resort to fiction to find counterfactual histories that will show their role. In *La Clef de Berlin et autre leçons d'un amateur de sciences*, Latour even illustrated the mediation of objects by using a comic book (CB: ch. 1). "Whatever solution is chosen, the field-work undertaken by ANT scholars has demonstrated that if objects are not studied it is not due to lack of data, but rather a lack of will" (RAS: 82).

Actor-network theory studies situations where it is still uncertain who or what an agent is and how it will co-exist with other agents. It deploys 'matters of concern' rather than 'matters of fact'. 'Matters of concern' is the expression Latour introduces to emphasize the uncertainty of what we are dealing with – an assembly, a 'gathering', a network that may show surprising action. *Networks* of translations do not transport causality but induce mediators into coexistence (RAS: 108). The translations generate the associations that make up the social. An ANT-account will describe (either human or nonhuman) actants as mediators participating in a string of actions, that is, being related to others through the translation they perform and through the translations that other actants perform on them. To do so, focus on the 'circulating object' in a chain of translations and abandon the idea that one has to deal with (given) 'matters of fact'.

ANT's use of the concept 'network' is a source of much misunderstanding. It should not to be confused with the concept of the 'network society' introduced by Castells (1996), nor with the conventional notion of a 'social network'. In ANT 'network' is a concept introduced to emphasize the role of chains of translations in making up the social. *It is a tool for description, not something out-there to be described.* An 'actor-network' is not a network of actors, but an assembly of actants who (by way of the translations they are involved in) are 'networked' and defined by the other actants. Latour's accounts of Pasteur, in *The Pasteurization of France* and in his analysis of Pasteur's *Mémoire*, exemplify the way in which *both* Pasteur and the microbes got identities by becoming entities in a network of translations. To understand what 'lactic yeast' refers to, we have to refer to Pasteur's actions; to understand what 'Pasteur' refers to we have to take his experiments with microbes and the hygienists into account.

Surely, sometimes in understanding how a collective is composed, we may find that actors have built a network in the traditional sense of the word. For example, they may have built what is conventionally called a 'social network' by collecting the names, telephone numbers or business cards of the people they have met at receptions or in meetings. But *where* is this 'social network'? We have to look into the address book in that person's mobile phone or ask him to hand over his stack of business cards. Anything that *exists* is set-up by associating heterogeneous elements. To describe how the GSM-network for mobile telephony is set up, one will have to describe the work of scientists and engineers, the negotiations between ICT-companies and governments about standards, to visit the offices of the European Union in Brussels and attend city council meetings where local sites for antennae are decided. In all cases, we are dealing with an assemblage of what conventionally are called 'technical' and 'social' elements. But ANT describes also what is not in common-sense terms conceived as 'a network' – e.g. a scientific paper – as a network of translations. To understand what a collective, a scientific paper, a technology or an organization is, ANT stresses the *work*, the movements, that went into bringing heterogeneous elements together, to deploy the various roles of (human and nonhuman) actants.

Similarly, the ANT-concept of *actor* is a strikingly different from the common sense notion of an acting person and from the concept of an actor in most of established sociology. We like to think that we are persons, conscious individuals who have cognitive abilities and who have reasons to act the way we do. We conceive ourselves to be the originator of our actions. This rather rosy picture had of course already been relativized by Marx, Freud and Durkheim. Yes we know that men make their own history, but that they do not make it under self-selected circumstances, but under circumstances existing already, given and transmitted from the past (Marx); we know that we are driven by unconscious drives (Freud); and even that in what appears to be the most individual act, committing suicide, the individual is subject to forces unknown to him as the regularity of social statistics shows that "[e]very society is predisposed to contribute a definitive quota of voluntary deaths" (Durkheim 1970 [1897]: 51). Marx, Freud and Durkheim pointed to 'hidden factors' not apparent in our common sense notion of an acting individual. As said before, there is nothing wrong with the idea of 'hidden' causes, provided that we take the trouble to specify *how* they make us act. However, often in social science "hidden variables have become packaged in such a way that there is no control window to check what is inside" (RAS: 50). So "we [. . .] have to increase the cost and the quality control on what counts as a hidden force" (RAS: 50). To force us to be more specific, the question to ask is what and who is doing the action and where the action takes place.

Where does individual action take place? Leave historical forces, unconscious drives and collective representations untouched. Just imagine a skipper steering his ship through difficult waters to reach a harbour. No doubt he is conscious of what he is doing. He uses a lot of equipment – compass, maps, a nautical slide rule, a radio to receive signals emitted by beacons, GPS perhaps – to calculate the safest route. Who is acting? Well, the skipper is in control, isn't he? He has made up his mind, he has planned the journey, he is the one who checks the readings of his instruments, he is doing the calculations, he is at the helm. When we concentrate on the *product* of his cognitive work – a safe journey to the harbour – the instruments appear as amplifying his cognitive abilities. Using them he can do what he could not do without these tools. But as Hutchins, a cognitive scientists and anthropologist, observed,

[w]hen we shift our focus to the *process* by which cognitive work is accomplished, [. . .] we see something quite different. [. . .] None of the component cognitive abilities has been amplified by the use of any of the tools. Rather, each tool presents the task to the user as a different sort of cognitive problem requiring a different set of cognitive abilities or a different organization of the same set of abilities. [. . .]

[These tools] do not stand between the user and the task. Rather, they stand with the user as resources used in the regulation of behaviour [. . .]. Rather than *amplify* the cognitive abilities [. . .], these tools *transform* the task the person has to do [. . .]. (Hutchins 1995: 154–155)

Who, what is the *source* of the action? Not only the skipper; his tools are (re)sources too. *Where* is the action performed? In, or rather by the network of translations that involve various nonhuman actants that transform the task of the skipper; action is *dislocated*. Where are the cognitive abilities of the skipper? As Latour already had discovered in Ivory Coast (cf. ch. 1), they should not be located in 'the mind', but have to be accounted for by connections. To find out who or what does what part of the acting, the ANT-analyst has to go to the details, rather than fall back on the common-sense idea that the skipper is acting according to the plans he made up in his mind. What he is doing is transformed by his instruments. They *make* him *do* something. The question who is active and in control and who is passive and controlled makes little sense. The French language has an expression – *'faire faire'*, meaning 'to make one do' and 'causing to be done' – that nicely points to the uncertainty about the authorship of action: yes, the skipper is active and his instruments help him in his tasks; but he is also controlled by what his instruments tell him – they make him do something. So, apart from being uncertain about groups, the ANT-analyst has also to be uncertain about action. Not only should he be prepared to allow nonhumans to act and to modify a state of affairs, he also should allow human action to be overtaken by other actants and to display the process, the network of translations, in which this is taking place.

Where does a private face-to-face encounter that the 'sociology of the social' often conceives as the most elementary form of 'social action' take place? For example, in a house. But as we have seen above, a private face-to-face interaction requires a secluded space, so that walls have to do something, namely preventing other people from intervening or overhearing what is said; and to be able to enter into the conversation, the door has to act predictably, as an intermediary, allowing us to enter the house. So who and what is acting? Surely, the two participants in the conversation, but also the mundane artefacts that allow them to have a private talk. And the latter would not be able to perform that task if builders had not done their work, the architect had forgotten to plan a door, and the local bank's account manager had refused the application for the mortgage to pay for the building. So even "face-to-face interactions should be taken [not as some concrete, primordial instantiation of social action, but] as the terminus point of a great number of agencies swarming towards them" (RAS: 196).

4.2   Deploying how the social is stabilized

The first task of the ANT-analyst is to be as undecided as possible about what makes up a group (an organization, an institution) and by who and what and where action is performed, to subsequently describe in detail 'the social', the movements (translations) that progressively make up an assembly, a collective, of heterogeneous – human and nonhuman – entities. However, 'the social' also stands for something else, namely for what has become stabilized and what presents itself as what Dahrendorf (1968: 50) aptly called "the annoying fact of society" – the fact that an individual can experience society as something given out-there that he may run up against like a brick wall. The mistake the sociologists of the social made, in Latour's view, was to mix up the two meanings of 'the social'. But once the two are clearly distinguished, there still remains the task of accounting also for the second meaning. So after having deployed the uncertainties of group formation and action, the ANT-analyst has to undertake a second task, namely to deploy the means through which 'the social' is stabilized and becomes 'society', what the sociologists of the social conceived as a set of 'macro' social structures, that presents itself as an 'annoying fact'.

One reason for becoming annoyed, we have already detected. Action is often overtaken. While we innocently perceive ourselves as the originator of our action, we find our actions to be transformed by others and by the instruments we use to perform our action. But how do we come to experience this as the effect of 'macro social structures'? Again, we have to be specific. *Where* are the structural effects actually being produced? We have to localize what appears as 'the global', the 'macro social world', *in* the local.

What appears as 'the macro' "is neither 'above' nor 'below' the interactions but *added* to them as *another* of their connections, feeding them and feeding off them" (RAS: 177).

How do actors 'macro-structure' reality (Law 1987; Callon and Latour 1981)? They do so by employing specific innovations. What are these innovations? In the first place *oligoptica* that allow for physically tracing connections elsewhere. 'Oligopticon' is the generic term for any site or device used to collect and process specific well-controlled information to allow "sturdy but extremely narrow views of the (connected) whole [. . .] – as long as connections hold" (RAS: 181). They are present in abundance. Cartographers collect and process information from all over the world to compose the maps that guide us in a foreign city or on the road (SA: 223–224; 236; PVI). In science, inscriptions from a wide variety of places are assembled in *centres of calculation*, to be compared and combined into manageable form and to be subsequently distributed and used elsewhere (SA: 232–257). 'Cameral sciences' (such as accounting, management, business organization) and statistical services provide similar means elsewhere (Porter 1995). They provide *forms, standards, metrologies* that allow connecting activities and sites by formatting translations and 'acting at a distance', to bring about effects far beyond the normal reach of humans. The CEO of McDonald's knows what goes on in the firm's franchisees in Singapore and Paris, because he will have the appropriate figures in a spreadsheet on his laptop and on that basis he can instruct his franchise-holders to act to his wishes. The CEO however also knows that to set up an oligopticon that provides these figures required a lot of work and that continuous maintenance is necessary to secure its proper functioning. And hopefully, he is aware that the spreadsheet on his laptop provides only a reduced picture of what goes on in McDonald's Singapore and Paris. But for his limited management purposes, it is enough for him to act upon. In ANT, the slogan is 'to follow the actors'. However we have to be aware that we not only have to follow them when they *multiply* entities, but also when they *rarify* them, by introducing oligoptica, replacing them by intermediaries, black boxes, and by standardizing them. We not only have to conceive actors as networks, but also to describe the (e.g. physical or administrative) networks and the oligoptica they added to extend their translations to be able to act at a distance.

Apart from oligoptica, also concepts like 'late capitalism', 'globalization' and the narratives that use such concepts provide a big picture, a *panorama*, that may help to orient us. Their nature, though, is different from oligoptica: "[w]hereas oligoptica are constantly revealing the fragility of their connections and their lack of control over what is being left in between their networks, panoramas give the impression of complete control over what is being surveyed, even though they are partially blind that nothing enters or leaves their walls except interested and baffled spectators" (RAS: 188). The sociologists of the social mistakenly took collective life to be composed on the basis of panoramas, where in fact it is performed on the basis of oligoptica. "At best, panoramas provide a prophetic preview of the collective, at worst they are a very poor substitute for it" (RAS: 189–190). They should be handled with care, or better: their use should be avoided.

Analyses of oligoptica provide the way to localize the global. But as the simple examples discussed before already show, 'the local' is a complicated notion as well. Even to understand the individual actions of the skipper, we had to take the instruments he used into account. To say that local (inter-) action is 'shaped' by many elements already in place doesn't tell us anything about the origin of those elements. Also the local will have to be re-dispatched and redistributed (RAS: 192).

Firstly, we may note that much is already in place. For example, the layout of a lecture hall suggests where the teacher will stand and where the students will be sitting. It provides a *script* for their actions. Moreover *plug-ins* may be provided: entering a supermarket, you will find the articles ordered on shelves by type of article and pre-packaged with information relating to content and price on the box. How could you ever make your very personal, intimate, rational choices if the articles for sale hadn't been displayed this way?

Secondly, we have to shift our attention from the actor to his *attachments*, that is to the outside connections that make him into an individual, a person. Cognitive competence is not just a mental, 'inside', capability, but also something that requires 'outside' connections, as both Hutchins and Latour's early work in Ivory Coast showed. Legal and official papers will designate you as someone with a specific name and identity. When you are romantically in love, you are guided by the films you watched, the novels you read and the songs to which you listened. They *subjectify* us. "[A]s William James [(1950 [1890])] so magnificently demonstrated, it is by multiplying the connections with the outside that there is some chance to grasp how the 'inside' is being furnished"

(RAS: 215–216). To understand the activity of subjects, their emotions and passions, we must turn our attention to their attachments that activate them and *make* them *do* something.

Once the uncertainty of action, the role of oligoptica, scripts, plug-ins and attachments is acknowledged, what the sociologists of the social cover by the very general term 'socialization' can – and should – be traced in detail. The vocabulary of ANT allows us to divert from the dichotomy between individual actor and society, from the dichotomy between actors being in control of their actions and actors being constrained in their freedom of action by 'social structures'. Action is not 'constrained' by plug-ins and attachments; on the contrary, they enable freedom of action. Try to live in a European city without legal papers – as illegal immigrants will testify, you will not have much freedom left; try to free yourself from the romantic images of films and novels and your next date will end in disaster. So "when we speak of an 'actor' we should always add the large network of attachments making it act. As to emancipation, it does not mean 'freed from [social] bonds, but [being] *well*-attached" (RAS: 217–218).

ANT introduces new terminology – like 'oligopticon', 'script', 'plug-in' and 'attachment' – to account for the common experience of 'the annoying fact of society', that is, the experience that individual actions and relations take place 'within' a 'society' (or alternatively within 'the economy' or 'a culture') that is perceived as a realm that guides and constrains individual actions. The new terminology may strike us as being cumbersome, counter-intuitive and superfluous. Do we really need all of that? Whatever defects it has, the 'sociology of the social' at least has the advantage of sticking closer to our everyday experiences of the social – on the 'micro-level' of social meaningful action and on the 'macro-level' of society. It codifies those common experiences in academic terms.

However, everyday experience is not a reliable guide and common sense may change when new forms of experience become available. Around 1600, the introduction of new instruments, especially the telescope, allowed astronomers to have new experiences. The discovery of Jupiter's moons compelled them to reconceptualize the Earth as one of the sun's planets – a notion that since then has become part of our common sense. If physics had rigidly stuck to everyday experiences, astronomy would still be based on the Ptolemaic system.

A similar shift may occur in what common sense takes to be 'the social' as the availability of new instruments, namely computers and the internet, allows us to experience the social in a new way (Latour et al. 2012). The professor who in the interlude halfway through *Reassembling the Social* didn't succeed in convincing the ambitious LSE-student to adopt ANT should have invited the student to spend an hour surfing the web and to reflect on what he was doing and experiencing. Clicking through platforms like LinkedIn.com or Academia.edu, going from webpage to webpage and from document to document, the student would have experienced that the web allows him to encounter people and to explore communities *without ever changing levels*. He would discover that on the web, the distinction between the 'micro-' and the 'macro-level' simply evaporates.

> If for instance we look on the web for the curriculum vitae of a scholar we have never heard of before, we will stumble on a list of items that are at first vague. Let's say that we have been just told that 'Hervé C.' is now 'professor of economics at Paris School of Management'. At the start of the search it is nothing more than a proper name. Then, we learn that he has a 'PhD from Penn University', 'has written on voting patterns among corporate stake holders', 'has demonstrated a theorem on the irrationality of aggregation', etc. If we go on through the list of attributes, the definition *will expand until paradoxically it will narrow down* to a more and more particular instance. Very quickly, [. . .] we will zero in on one name and one name only, for the unique solution: 'Hervé C.'. Who is *this* actor? Answer: *this* network. What was at first a meaningless string of words with no content, a mere dot, now possesses a content, an interior, that is, a network summarized by one now fully specified proper name. The set of attributes – the network – may now be grasped as an *envelope* – the actor – that encapsulates its content in one shorthand notation. (Latour et al. 2012: 592–593)

The identity of the actor has become defined by following a network of connections. By following a string of links we have 'individualized' him. Reversibly the network is fully defined by its actors. If, for example, we want to know what this strange university, the 'Paris School of Management', is, the list of its academics on its website provides the answer. "By circulating [. . .] from the actor to the network and back, we are not changing levels [that is, moving between the micro-level of the individual and the macro-level of the organization he works 'in'] but simply [are] *stopping momentarily* at a point, the actor, before *moving on* to the attributes that define them" (Latour et al. 2012: 593). The digital environment of the web allows us to experience 'the social' in a new and by now quite common way, by following actors through their connections, that is, in ANT-style.

The web was not available to Weber and Durkheim. Given the tools available to them, they may be excused for conceptualizing 'the social' on two levels, pointing to on the one hand the experience of meaningful action and on the other hand the statistical data about patterns of behaviour and aggregate facts. However, when Tarde's works were republished (e.g. Tarde 1999 [1895]), Latour and other ANT-scholars discovered that this French jurist, criminologist, philosopher and early sociologist had already anticipated much of what they had developed under the name of ANT. Also Tarde simply lacked the appropriate instruments to fully develop his ideas; he had to articulate them in abstract concepts (as 'monad', borrowed from Leibniz) and metaphors. Having been defeated by Durkheim in heated discussions about the nature of sociology, his work had almost completely been forgotten. But navigating through digital datasets allowed Latour et al. (2012) to explore and test his ideas. In Tarde, ANT found its pedigree (Latour 2002; RAS: 13–15), the ancestry any serious intellectual discipline needs.

By navigating through the by now abundantly available digital datasets, we may experience 'the social' in a new way. However, it also shows what is badly needed: more advanced tools to convincingly visualize actor-networks and their dynamics – tools that will allow easy manipulation and transfer of ANT-analyses. As Latour's early studies of laboratory science showed, to really blossom no discipline can do without them (Latour 1990b). Latour has taken that lesson to heart. To explore fully the potential of ANT and "to render the social sciences empirical and quantitative without losing their necessary stress on particulars" (Latour et al. 2012: 613), he initiated a 'medialab' at Sciences-Po (www.medialab.sciences-po.fr), to develop new digital tools.

4.3   Shifting focus

Actor-networks are assemblies of human and nonhuman actants, Latour emphasized both in his work in science and technology studies and in *Reassembling the Social*. But in the chains of an actor-network set up by assembling heterogeneous elements to serve some specific function, something circulates or is passed on.

ANT-analyses can be focused either on the *set up* of an actor-network, or on what it allows to *circulate* once everything is in place.

Latour illustrates the difference with a simple example. To describe how the network that runs from Russia into Europe to deliver natural gas is *set up*, one will have to talk about big pipes, pumping stations, valves, technicians, business contracts, international treatises and Russian Mafiosi – stuff for "a real John le Carré novel". But what is *passed* through this network is none of the former, but natural gas. "So under the word 'network' we must be careful not to confuse what circulates *once everything is in place* with the *setups* involving the heterogeneous set of elements that allow circulation to occur" (AIME: 32).

In *Reassembling the Social*, the heterogeneity of the elements that set up actor-networks was emphasized. Sociologists had to learn that 'the social' is not some specific realm, but the name for the process of assembling a collective out of heterogeneous elements. But once set up, actor-networks may serve different functions. Science, law and politics are different institutions. A laboratory, a court of law and a parliament have different functions. But in all of them, we will find people grappling with papers, books, and emails. Of course, there are differences – one would not expect to encounter Erlenmeyer flasks in a court of law. But a scientist may appear there – either as defendant, or as an expert witness – and on the desk of a Member of Parliament reports written by economists and scientists may lie. How may one *empirically* contrast various institutions? Latour suggests: by focusing on what *circulates* in the actor-networks that are set up for them to function, that is, to serve a particular value. In *The Making of Law* (2010, French edition 2002) he explicitly set out to undertake that task.

Having spent two decades studying science and technology, in the early 1990s Latour decided to study law. He obtained permission to attend both public and some of the non-public sessions of the *Conseil d'État* to do an ethnography of the French High Court for administrative law. The book that came out of this research is *The Making of Law*.

Established in 1799 by Napoleon Bonaparte and based in the *Palais Royal* in Paris, the *Conseil d'État* has two functions. In the first place, the Council advises on the legal quality of bills and amendments to laws that the French government is about to propose to parliament. Its second function is to serve as the final court of appeal for legal conflicts between citizens (or other legal persons, e.g. companies) and the state. Its decisions are precedents-based only; in contrast to its statutory criminal and civil law, in France administrative law is common law.

As the appeals discussed in *The Making of Law* make clear, administrative law may relate to almost any aspect of society. One of the cases Latour discusses at length concerns a farmer whose crops have been damaged by a town's pigeons and who has filed an appeal after having tried unsuccessfully at lower courts to obtain compensation from a commune for the damage; another case deals with a company objecting to the appointment, signed by the President of the French Republic, of a former civil servant who changed his position as a civil servant overseeing a bank to become the president of that bank; a third example concerns an appeal against the state's decision to expel an Iraqi asylum seeker with a criminal record. The law touches everything, but not everything is the law. The task Latour takes up in *The Making of Law* is to describe what makes law specific.

The Making of Law is based on fifteen months of observations conducted in the mid 1990s. The similarity with Latour's study at the Salk Institute soon becomes clear: again we enter a small world of people meticulously discussing texts at length. In contrast to a scientific laboratory, the Council uses no sophisticated instruments, nor does it engage in fieldwork. All cases are exclusively discussed on the basis of the documents an appellant's lawyer has submitted and on the books and databases that contain precedents, the decisions the Council has taken in earlier cases.

Leafing through *The Making of Law* may make one wonder whether actor-network theory has played a role in Latour's enquiry at all. An ANT-ethnographer should go where his study will lead him. To study scientific controversies, ANT-scholars have moved from laboratories to parliaments, city halls and corporate boardrooms. But in his ethnography of the *Conseil d'État*, Latour stayed in one place, namely the richly decorated rooms of the *Palais Royal*. He didn't follow any links outside its walls, e.g. to see the effects the Council's decisions bring about. Oligopticas, plug-ins, scale or other technical terms from ANT are absent from his account. Granted, by pointing to the pecking order in which the various participants are seated in conference rooms and to the way the Council members' letterboxes are lined-up according to seniority, and by discussing some of the physical aspects of handling files, Latour addresses some of the 'material aspects' of the passage of law. But is that sufficient to call an ethnographic account an exercise in ANT?

Also some individual chapters may make one wonder whether we are dealing with ANT-ethnography at all. One chapter might have found a place in any treatise in the sociology of professions. It deals with the biographies and career-paths of counsellors. Latour finds the majority of members of the Council to have graduated from ENA, the elite National School of Administration. But not all of them have been trained in law: one-third of the counsellors comes 'from the outside' – they have made careers in journalism, politics, government, the military, and medicine or (occasionally) in science. After some time, many counsellors will find employment outside the legal system, in either the public or the private sector. Any 'sociologist of the social' could have written that.

But first impressions are deceptive. On the last page of the chapter that maps the career-paths of counsellors, Latour shows his true colours. The chapter has shown that counsellors float in and out of the Council and that their backgrounds vary. So, Latour concludes, it would be quite wrong to conceive the law *as a separate sphere* or as well-delineated subsystem of society. However, immediately after having concluded this, Latour adds: "And yet *it is*" (ML: 126). The fabric of law establishes a *specific* type of association. To 'reassemble' this type of association, Latour's sets out to describe ethnographically how and under which conditions this highly specific type of association is attained. Instead of describing what it takes to *set up* the Council – counsellors with various backgrounds who move in and out, stacks of documents, file folders, letterboxes and office paraphernalia – he decides to focus on what is *passed* through this heterogeneous network.

For this purpose, he does not follow the counsellors, but the objects circulating in the long chain of translations that make up the passage of law: files. He substitutes "the grand talk about Law, Justice and Norms with a meticulous enquiry about files – grey, beige or yellow, thin or thick, easy or complex, old or new [. . .]" (ML: 71). To see what Latour is after, we too will have to follow these files from the point where an appeal from a litigant's lawyer has arrived at the *Palais Royal* up to the point where the Council of State has reached a decision either to annul the decision of a lower court or a state authority, or to reject the litigant's claim.

Before an appeal arrives by mail at the *Palais Royal*, lower courts will already have dealt with a litigant's claim, but not to his satisfaction. So he has decided to appeal. Citing facts, laws and decrees, the appellant's lawyer will have displayed in detail the reasons for the claim that so far justice has not been done to his client. In the

Palais Royal, the lawyer's letter and the documents he has sent to backup his claim are stored in a file-folder.

The folder is subsequently passed to the Analysis service. This office examines the appeal and assesses its character; it adds a list of relevant precedents to the file and starts to fill the folder by sending out requests for submitting any missing documents (such as confirmations of powers, communication of the [case in] first instance, etc.) (ML: 77). After its passage through the Analysis service, the case possesses a number, a coloured folder (specific for the type of case) and a form that classifies the file and that specifies the stage of proceedings. The file folder (literally) becomes weightier. The folder then goes to the sub-section of the Council that is responsible for the specific type of case. The defendant is given the opportunity to reply to what is by now in the folder; the defendant's reply is also added to the folder. In the next step of the procedure, the file is handed over to a specific official, called the reporter.

The reporter's task is to extract *moyens de droit* from the pile of papers that by now constitutes the file, in order to link them to other texts. A *'moyen'* – a legal term difficult to translate into English (ML: x) – is a legal ground, a reason, an argument on which legal judgment can be based. To extract the *moyens*, the reporter will establish a *relation* between two types of texts: on the one hand the various documents in the file folder and on the other hand legal texts that are "filed, established, accepted, referenced and assured" (ML: 86) in one way or another – the record of 200 years of French history of administrative law, its precedents and commentary, available in volumes on bookshelves and in databases. Linking the two types of documents, the reporter is (metaphorically) building a bridge on which

> *something* is now going to be able to *pass* from one side to the other. Either in one direction, and this is annulment, or in the other, and that is rejection. [. . .] Either the claim will be given the power to follow its course and break one of these texts – orders, decisions, regulations or decrees – which together form the immense cloth of published texts: this is annulment. Or conversely, there is something in the published texts which has such force that it transports itself from one side to the other and blocks the progression of the claim definitively: that is rejection. (ML: 88–9)

The reporter will discuss the relation between the documents the appellant's lawyer has filed and the legal texts and precedents and he will formulate on this basis a conclusion: either the earlier decision of a lower court should be annulled or the litigant's appeal should be rejected.

Has final judgment been reached? No. The procedure at the Council of State does not stop after the reporter has finished writing up his conclusion. So Latour follows the file on the next leg of its journey. The reporter's note and the file-folder are handed over to another person, the 'reviser'. He will summarize the entire file and test the weaving work done by the reporter. He will subsequently present the string of arguments orally in a review session, a meeting that includes another person, the 'commissioner of the law'.

The commissioner of the law is confronted with the appeal at hand during the review session for the first time. In most cases, he will just listen to the reviser's oral presentation and the discussion it may invoke. A few weeks later, the commissioner of the law will write up *his* conclusions. Some time – weeks, months – later, he will read his conclusion in a public court session in front of the judges. The judges will subsequently retire to discuss the case among themselves and in the silent presence of the commissioner of the law and the reporter, in order to decide whether or not they follow the conclusion. Again, some months later, all concerned parties in the conflict will be informed about the decision. The judgment is also displayed on a noticeboard at the foot of one of the *Palais Royal*'s staircases. By then, the case is closed definitely.

What is the point of this strange procedure in which time and again the weaving of texts that the reporter has achieved is reviewed, discussed, taken up two times by other officials, after which the commissioner of the law's conclusions are read in court, to be discussed, once again, by the judges before a final decision is made and announced?

Tracing links is the core-business of ANT. But here we find a practice where the weaving of texts and establishing links is interrupted time and again to allow a fresh look at what has been done so far and a new round of discussions. Of course, an ethnographer cannot neglect this. He needs to have a closer look at this strange procedure that seems to indicate that the Council explicitly hesitates, before it decides to finally do its job of 'saying the law' (*'dire le droit'*) and to administer justice.

Latour decides to take a closer look at the counsellors' and judges' reasoning. Of course, their internal mental processes are not accessible for an ethnographer. In fact, 'reasoning' is an ill-chosen word. Someone who is 'reasoning' is oriented towards an outcome; he intends to draw a conclusion. Latour decides not to focus on what the counsellors intend to do, but to shift his attention to what they are doing and what *makes* them *do* something. In the same way that in Hutchin's study discussed earlier navigation instruments make the skipper do something, he finds the counsellors

> grappl[ing] with a file which acts upon them, which pushes and forces them, and which *makes them do something*. Nothing gives a greater impression of resistance, or of being a thing or a cause. But at the same time this material has a very particular plasticity because each agent [. . .] modifies the form taken by the arguments, the salience of texts, and traces on this ectoplasm of the administrative law a set of divergent paths, mobilizing clans who confront each other with facts, precedents, understandings, opportunities or public morality, all of which are used to stoke the fire of the debate. (ML: 192 italics added)

To try to grasp what in this process is *passed*, Latour sets out to detect "explicit signs of the changes of position of the members of the Council they make with respect to the files that they are dealing with" (ML: 129). What should these signs indicate? The explanation Latour provides is that they should show "the transition, movement or metamorphosis of the particular force whose dynamic we are attempting to reconstitute" (ML: 129). Latour has written clearer sentences. To understand what is meant, we need to step back first. *Reculer pour mieux sauter.*

In *Irreductions*, Latour wrote: "We neither think nor reason. Rather, we *work* on fragile materials – texts, inscriptions, traces, or paints – with other people. These materials are associated or dissociated by courage and effort; they have no meaning, value, or coherence outside the narrow network that holds them together for a time" (IRR: 2.5.4). What Latour sets out to detect is how and under what conditions the materials (the files) get "meaning, value, or coherence", that is, legal binding quality, in the network of translations that is established by moving the files through the long and often interrupted process at the Council.

Not every move will give the appropriate "meaning, value, or coherence" to materials. In football, the ball is the 'circulating object'. If a player kicks a lump of clay into the goal, he has not scored; nor has he scored if he has moved the ball by his hand or when his position in the field was 'offside'. In football, rules apply.

They are listed in the laws of the game. A player scores a goal only when the object he has moved carries a *specific value* (when it is the official ball with which the game is played) and when the action is performed under *specific conditions*.

So do we have to consult some textbook of law to find the rules that apply in the Council of State? No, an ethnographer has to detect the value object and the conditions that apply on the basis of his observations. To perform the latter task, Latour shifts from semiotics to the theory of speech acts, a sub-discipline of philosophy of language inaugurated by Austin (1976 [1962]).

Before Austin, most philosophers of language focused on the relation between statements and the states of affair they describe, that is, the facts that they state either truly or falsely. Austin pointed out that many sentences do not state a fact, and are neither true nor false, but that in uttering these sentences the speaker performs some act. The speaker 'is doing things with words'. For example, by uttering his words the speaker is giving a warning, advising, or making a promise. Austin refers to such actions performed by saying or writing something as 'illocutionary acts'. He subsequently asked under which conditions an utterance has 'illo-cutionary force', that is: the force to change a situation in a specific way. For example, when Peter, addressing Mary, utters the words 'I promise to pay you €100 tomorrow', he is doing something: he is giving a promise to Mary. By uttering these words, he is chang-ing the situation. A semiotician may analyse what has happened in terms of the three actantial roles introduced in § 2.3: an opera-tive subject (Peter), a passive subject (Mary) and an object (the words Peter has spoken). But to specify *in what way* the situation has been changed by these words we need something else; we need to specify the 'illocutionary force' of Peter's utterance. By uttering his words, Peter has made a promise, rather than given an advice or a warning. So the question becomes: what are – in the terminology that Austin introduced – the *conditions of felicity and infelicity* for veritably making a promise (rather than giving an advise, or a warning)? If we think about this for a minute, we will note that various conditions have to be met. For example, Peter has to speak his words in a language Mary understands. But also where and when the words are uttered matters. If Peter has uttered his words to Mary on stage in a play, he would not be committed to keeping his promise after the curtain has fallen – he surely would be surprised to find his co-star the next morning on his doorstep to collect the money. If the conditions of felicity are not fulfilled, an utterance is 'void', without illocutionary force. The utterance does not point to what conventionally, under its proper conditions of felicity, would follow: a *specific* change of state in which Peter has committed himself to giving €100 to Mary the next day; and in which Mary has become the future recipient of that money.

So, to understand what is going on in the Council, Latour, the ethnographer, not only has to follow files. To understand how, by moving these files through subsequent meetings and discussions, 'meaning, value, or coherence' is produced, he also has *another* task: he has to extract both the 'value object' and the 'conditions of felicity and infelicity' of legal talk and action. This is what Latour meant in the rather unclear passage quoted above in which he announced that "he will try to grasp what the counsellors make *pass* through their interactions which allow them to judge the quality of their action" (ML: 129). He sets out to (empirically) detect signs that point to the *value* of the circulating object and to the *conditions of felicity and infelicity* of *'dire le droit'* ('saying the law'). Only when the right value object is passed and the right conditions of felicity apply, will the decision of the Council of State have legal, binding force on the parties of the litigation. By extracting the value object and the conditions of felicity and infelicity from the signs, Latour sets out to describe law as a specific 'regime of enunciation', that is, as a practice of speaking and acting that has its own *normativity*.

So are we, after a long and winding road, back on the terrain of traditional philosophy, and should we start discussing the compelling normative force of moral or – in this case – legal reasoning? No. As we have seen before, to say that the members of the Council of State 'reason' is to shortcut what has to be described in empirical terms. They grapple with texts, they confront each other with "facts, precedents, understandings, opportunities or public morality, all of which are used to stoke the fire of the debate."

And when this process comes to an end, it is never because pure law has triumphed, but because of the internal properties of these relations of force or these conflicts between heterogeneous multiplicities, and because the actors themselves consider that certain value objects have indeed been transferred and that conditions of felicity have indeed been fulfilled. (ML: 192)

So is Latour trying to detect precisely what the sociologists of the social referred to when they refer to 'collective representations', the norms and values of institutions, organizational cultures, mentalities and symbolic structures to explain why actors are constrained in their action? No. The difference is both subtle and wide. By extracting the 'value object' that the actors themselves think need transferring and the 'conditions of felicity' that have to be fulfilled, Latour tries to detect the *internal properties* of the relations that are established in the Council's practice. He is not attempting to 'explain' the action of counsellors by norms, values, power or social structures that guide or constrain their actions from the outside. He is interested in describing *what* is done, rather than in explaining *why* it is done.

In a long discussion (ML: 129–197), Latour extracts the 'value object' and the 'conditions of felicity' from the discussions in the Council. Fusing language from semiotics and Austin's philosophy of language, in *The Making of Law* 'value objects' and 'conditions of felicity' are lumped together (e.g. summarized as 'value objects' (ML: 194–195), which five lines later are nevertheless called 'conditions of felicity'). When Latour comes back to law in *An Inquiry into Modes of Existence*, a more refined language has been introduced and the two become separated again (cf. ch. 6). For now, let's stick to the distinction as introduced before, using the analogy of football.

The 'value object', we have already met: the *moyen*, the legal reason, ground, or argument the reporter extracts from a file to relate its text to laws, decrees, and precedents, or any other previously established legal text. This is what has to *pass* (compare: the official ball) to give the final decision of the Council legal force (compare: 'to score a goal'). But what is passed on will only have this illocutionary (i.e. legal binding) force if the proper conditions of felicity for 'saying the law' are met (compare: 'the ball was not kicked in the goal by a player who had been off-side').

What are these conditions of felicity? A first one is deduced from the frequent explicit interruptions of the process, to give a case a fresh look. What's its point? The interruptions build *hesitation* into the process, to produce freedom of judgment by unlinking things before they are linked up again; they further the *quality* of the process. Two other items on Latour's list are the *coherence* of law itself as it modifies its internal structure and quality (which shows for example in the care the Council takes in presenting any reversal of precedents as an improvement of their predecessors' original intent); and the *limits* of the law, which are defined by regulating the right to launch or suspend a legal action. Three conditions relate to the process itself: the *progress* of the claim as it moves through obstacles and delays is carefully watched in the understanding that eventually a final decision has to be made; as are the *organization* of cases, which enables the logistics of the claim to be respected without too many mistakes and the process of *quality control* by means of which the conditions of felicity of the process are verified reflexively, e.g. by asking questions like 'have we judged well?' and 'is the discussion closed?' Finally, there is the *interest* of cases "without which the counsellors would long since have died of boredom" (ML: 191); the *authority* of the agents participating in the judgment which undergoes a meta-morphosis throughout the ordeal of the process of review; and the *weight* of texts, the authority of the legal objects rather than the human subjects, modified by appealing to precedents, or codes. Together they constitute conditions of felicity for speaking law truthfully and for bringing things to an end, that is, for 'saying the law'.

To *set up* the actor-network that makes up the *Conceil d'État* requires a heterogeneous set of elements: professionally trained counsellors, office equipment, a library, databases, meeting rooms, et cetera. But to describe in what way this network functions in a specific way, namely as a court of law rather than – say – a scientific laboratory, Latour has turned his attention to what is *passed* in this network. And by pointing to the value object and the conditions of felicity that apply, he has detected how *in* this actor-network a special kind of connection is constituted, namely legal ties. Only if the actor-network as a whole functions well will the final judg-ments of the Council have the legal binding force to either annul a decision of a lower authority, or to reject the appeal that an appel-lant's lawyer has filed.

Having studied one peculiar form of law (precedents-based administrative law, not statutory criminal or civil law) in one par-ticular French institution, Latour draws two conclusions that go beyond this limited case, one relating to sociology, the other to philosophy of law.

In the first place, he rejects the idea that the force of law can be explained in sociological terms by pointing to what is 'behind' the law in the 'social context' – power, interests, or social structures of whatever kind. His reasoning is familiar; he used it also in criticizing the sociology of scientific knowledge.

> What is true of the sciences is even more so of law: how can law be explained in terms of the influences of the social context, when law itself secretes an original form of contextual networking of people, acts and texts, so that it would be very difficult to define the notion of social context without resorting to legal concepts? There is no stronger metalanguage to explain law than the language of law itself. Or, more precisely, law is *itself its own metalanguage*. (ML: 259–260)

Law is mixed with everything, but in a specific way; "it judiciarizes all of society, which it grasps *as a whole* in its own *peculiar* fashion" (ML: 262). To make this peculiar fashion visible, Latour has "extracted the legal work from the institution like a physiologist might have extracted bone marrow from a dog, knowing full well that it is not the entire animal. To study the Council of State, [he has] disregarded the State – and willingly acknowledge[s] the paradox" (ML: 253).

But if neither the State, nor hidden social structures explain the force of the law, where does this force originate? First note that this question is a different one from: how is this force played out? To enforce the law outside the *Palais Royal* – or any other court of justice – of course requires much more than discussions among counsellors and judges under the right conditions of felicity. To enforce the law, to make the force of law "durable", requires organizations, technologies, and if necessary arms and jails (McGee 2014: 165 ff.). But that is not where we will find the normative – legally binding – force of the law. To think otherwise would erase the distinction between societies where people are put behind bars without having due process and societies that respect the rule (and thus the normative force) of law.

To answer the question where the force of law originates, the tradition of the philosophy of law has two prominent answers available. According to the first one, the 'natural law' (*jus naturalis*) tradition, the force of law is determined by nature; in this tradition, the basic principles of the law are conceived as basic reasons for action and basic human goods. The second answer, 'legal positivism', understands the law entirely as a human product, as the aggregate of – in the history of law agreed upon – conventions, of what is called 'positive law'. The similarity with the contrast between realism and social constructivism in the discussions about science is apparent. Given that Latour rejects both positions with regard to science, it will not be surprising that he also rejects both natural law and legal positivism. In the final chapter of *The Making of Law*, he draws a second general conclusion from his casestudy, a philosophical one.

"To celebrate [the value of law], there is no point in playing the great organs of nature, religion, the State. Law already has enough totality attached to it for us not to add all these dead weights that, in any case, make an awful noise" (ML: 277). So is legal positivism right, is law 'just a human construction', a matter of contingent conventions as legal positivists claim? No, not quite either. For sure, human decisions are involved and there is a lot of contingency. Aggregation of precedents is not based on some rational plan. Prior decisions of the Council have entered the books like shells washing ashore. But this is no reason to belittle their role. What has started contingently becomes binding necessity. The Council's judges accept the prior decisions as being *correct* ones. They set a norm, they define what is and what is not a *moyen*. The prior decisions are taken as articulations of what the Law says. If a precedent is found to be out-dated and has to be reversed, by a Whiggish rewriting of history, arguing that the old text should be interpreted not as it was written but as the authors would have reacted had they had before them the same facts as today, the judges make the tradition to be coherent and to have been coherent (ML: 233 n.42).

So law is a human construction after all? Yes and no. Latour's position on law is similar to the one he argued for with regard to science: constructivism and realism are not incompatible (cf. § 2.4). Law is a human construction, but law is also real, it defines us as human. "Without [law], we wouldn't be human; without it we would have lost the trace of what we had said. [. . .] We would be unable to find the trace of our actions. There would be no accountability" (ML: 277). Law is based neither on 'nature', nor on 'human conventions'. The force of law originates *within* law. That is meant by the phrase that "law is *itself its own metalanguage*'. Law has its own *ontology* that – as the ethnography of the *Conseil d'État* has made apparent – not only involves humans, texts, file-folders and paperclips, but also specific value objects and conditions of felicity. Introducing a term that would occupy him at length in *An Inquiry into Modes of Existence* but that in *The Making of Law* is left unexplained, Latour formulates as his conclusion: law is a particular *mode of existence* (ML: 247).

By not only focusing on how the actor-networks that make up the *Conseil d'État* is *set up*, but also by discussing at length what is *passed* on in this actor-network, Latour has detected what is specific for law as an institution. It has its own 'mode of existence'. Fleshing out the characteristics of modes of existence in general terms, will allow him to specify what is specific for other institutions, like science and politics, to better *contrast* them. We'll get to that in Chapter 6.

5

A Philosophy for Our Time

The experience with analysing science and technology not only led Latour to suggest an alternative social science. He used the conceptual apparatus that came out of his work in science and technology studies also to address the issues that increasingly troubled citizens, politicians, and scholars: ecological problems, the role of science in democracy, and ultimately the way we understand ourselves as modern people – or rather 'postmoderns', as some claim. Latour set out "to bring the emerging field of science studies to the attention of the literate public through the philosophy associated with this domain" (NBM: ix).

It's the late-1980s. In Eastern Europe, unrest spreads, which eventually will lead to the fall of the Berlin wall. Meanwhile, at international conferences, new global issues are discussed – the hole in the ozone layer, climate change, the depletion of the planet's natural resources and raw materials. The world becomes aware that it is facing new threats – threats, the causes of which transgress national borders, while their projected effects will affect future generations, and that thus exceed both the jurisdiction and the time-horizon of nation-based politics. Observing that environmental, economic and social threats increasingly tend to escape the established institutions for monitoring and protection, the German sociologist Beck (1986) announces that modern, industrial society has entered a new phase. He characterizes the transition as the advent of the 'risk society'.

The new threats do raise questions about the role of science in democracy. Identification of the new threats depends on scientific knowledge. Citizens can detect stench themselves and their protests may urge politicians to take action; but that political action is required because exposure to some chemical compound may cause cancer in the decades ahead and because current levels of CO_2-emissions will eventually cause climate change with disastrous effects, is known only because of the work scientists have invested in the matter. So scientists no longer restrict themselves to what the idea of the role 'value-free science' in politics suggests, that is, to provide advice about the means to attain ends that democratic governments have decided to pursue. Increasingly their work defines the political agenda. However, the experts disagree on many issues, leaving citizens and politicians in the dark about what to think and what to do. How should a democracy deal with risks, uncertainty and the new role of science? The risk society calls for a "reinvention of politics," Beck (1993; Beck, Giddens and Lash 1994: 1) argued.

Appraising Beck's social theory as "one of the most lively, creative and politically relevant forms of sociology developed in recent years," Latour (2003) nevertheless thinks that Beck has misunderstood the problems we are confronted with. Beck thinks that the problems emerged because society has changed; according to Latour the crisis of modernity is even more profound: it shows that we cannot any longer live with our *self-image* as Moderns.

What is this self-image? For most of the twentieth century, the answer had seemed pretty clear. We – the peoples of Europe and North America – self-consciously declared ourselves to be Moderns. Modern, that is: to be no longer bound to the social practices and values of traditional societies, to be enlightened; in Kant's famous words: to be released from man's self-incurred tutelage. Science enables us to rationally control nature; democracy and the rule of law guarantee our political freedom. They are two of the keystones that define societies as 'modern' ones.

All diagnoses of modern society and culture are heirs to the analyses of Durkheim and Weber, the classical sociologists of the early twentieth century who tried to capture the nature of industrial societies that had developed in the century before. Weber, in particular, defined the terms in which for decades the debates about 'modernity' would be conducted.

Weber's extensive, comparative studies of European and non-European religions, economies and legal practices, led him to conclude that Europe and North America had developed distinctive features in three domains: culture, society and personality (Weber

1968 [1919]; 1972a [1922]; 1972b [1922]). For the domain of culture, he pointed to the development of modern science and technology, formal law, principles-based ethics and Western art, in particular harmonic chord music. For the societal domain, he cited the development of capitalism and rule-based, bureaucratic government. Finally, he characterized the modern West as having developed the personality of the professional, distinguished by its calculative attitude that allows strict separation between matters of business and personal concerns. This wide variety of social developments has a common denominator, Weber argued. All of them rely on methods, procedures, on calculability and predictability. What according to Weber makes the modern Occident exceptional in both time and space is its particular form of rationality, namely calculative, purposive-instrumental rationality, *Zweckrationalität*. He conceived Western rationalism as the product of a long process of intellectualization and rationalization.

Being modern means having learned to distinguish between 'ought' and 'is', between mind and matter, between on the one hand society and culture – the sphere of values – and on the other hand nature – the sphere of brute facts. Moderns know how to distinguish ends from means and they know that ends do not follow from facts, but are chosen on the basis of values. But, Weber stressed,

> [t]he increasing intellectualization and rationalization do not [. . .] indicate an increased and general knowledge of the conditions under which one lives. It means something else, namely, the knowledge or belief that if one but wished one could learn it at any time. Hence, it means that principally there are no mysterious incalculable forces that come into play, but rather that one can – in principle – master all things by calculation. This means that the world is disenchanted. One need no longer have recourse to magical means in order to master or implore the spirits, as did the savage, for whom such mysterious powers existed. Technical means and calculations perform the service. This above all is what intellectualization means. (Weber 1968 [1919]: 593–4)

To understand and explain what set off the process of economic rationalization, Weber pointed to the sermons that helped to spread the asceticism of the Protestant work ethic (Weber 1972c [1922]). But "the most important fraction of the process of intellectualization" had started already earlier. Weber situates its start in Plato's allegory of the cave, in the discovery of the significance of knowledge and understanding that opened "the way for knowing and for teaching how to act rightly in life and, above all, how to act as a citizen of the state" and in the appearance of the rational experiment – "the means of reliably controlling experience" – during the Renaissance period (Weber 1968 [1919]: 596).

Rationalization has brought indubitable advantages, but Weber also acknowledged its downsides. As bureaucratic rationalism may become an "iron cage" from which there is no escape, it meant "loss of freedom". Moreover, disenchantment of the world means "loss of meaning". Swammerdam, a seventeenth-century Dutch scholar, could still perceive the proof of God's providence in the anatomy of a louse. But "[w]ho – aside from a few big children who are still found in the natural sciences – still believes that the findings of astronomy, biology, physics, or chemistry could teach us anything about the 'meaning' of the world?" Weber (1968 [1919]: 597) rhetorically asked. For rationalized people, the gods have deserted the world.

Weber's analyses gave sociologists and social philosophers a bone to chew on for many decades. Elaborating worries about 'loss of meaning' and 'loss of freedom', the Frankfurt School, up to Habermas (1981), turned Weber's concerns into an influential research programme. Identifying rationality and reason with enlightenment, the Frankfurt philosophers wondered why rationalism had resulted in capitalist exploitation, bureaucracy's iron cage, and 'one-dimensional man', rather than in freedom and emancipation. Having first declared reason to be an almighty god, they subsequently had to find an explanation for the evil that existed in rationalized societies. They had to address the secular version of the question of theodicy.

Weber's analysis began to meet more criticism only in the last decades of the twentieth century. Anthropologists showed other cultures to have intricate forms of rationality of their own. For example, trying to make sense of complicated legal procedures on Bali, Geertz (1983: 179) ironically observed: "we have here events, rules, politics, customs, beliefs, sentiments, symbols, procedures, and metaphysics put together in so unfamiliar and ingenious a way as to make any mere contrast of 'is' and 'ought' seem – how shall I put it? – primitive." Kuhn's critique of the idea of a unified scientific method put the whole idea of an overarching concept of rationality guiding practices in question (Hollis and Lukes (eds.) 1982). If an appeal to rationality couldn't do much work for explaining scientific development, the prospects for the concept of rationality to perform the much more encompassing role Weber attributed to it were dim. Under the broad label of 'postmodernism', disappointed rationalists started to express scepticism about all 'metanarratives' – whether they related to politics, science, or the arts. But Weber is hard to surpass. As long as discussions about the nature of modern society take the concept of rationality as their battleground, they are conducted in his shadow.

That it is hard to get beyond Weber is little surprising. Indeed, it is undeniable that what he singled out as exceptional Occidental achievements do still characterize modern society: capitalism, science, formal law, bureaucracy, and the calculative attitude of modern professionals. So it may come as a surprise that Latour (NBM) claims: yes, indeed, we have all of that, but *we have never been modern*.

5.1   'We Have Never Been Modern'

Again, the analysis starts by calling attention to a text – this time not a text published in a scientific journal, but a newspaper article from *Le Monde*. Latour's daily reports that measurements taken above the Antarctic show the hole in the ozone layer to have grown ominously larger. Because the ozone layer prevents ultraviolet rays from harming plants and animals on Earth's surface, this is alarming news. A few lines later, the article shifts to CEOs of big companies who will modify their production lines to replace the chlorofluorocarbons that have been identified as the cause of the depletion of ozone in the stratosphere. The article continues to report that heads of state are involved in debates about chemistry, refrigerators, aerosols and inert gases too. However, the paper also says that meteorologists suggest that cyclical fluctuations rather than human activities account for the observed changes in the ozone layer. So both industry and governments no longer know what to do. Finally, towards the bottom of the page, the reader is informed that Third World countries and ecologists are talking about international treaties, moratoriums, the rights of future generations and the right to development. One and the same article

mixes chemical reactions with political reactions, links the most esoteric sciences and the most sordid politics, the most distant sky and some factory in the Lyon suburbs, dangers on a global scale and the impending local elections or the next board meeting. The horizons, the stakes, the time frames, the actors – none of these is commensurable, yet there they are, caught up in the same story. (NBM: 1)

Expecting science, politics and economy to be clearly separated domains, with each of them having their own institutions, methods and procedures, any reader of Weber would be surprised to find a report in a serious newspaper that mixes facts and values, knowledge and doubt, human activities and ozone high up in the stratosphere, science, technology, politics and economy. But not Latour. Having been involved in science and technology studies by then for more than a decade, he is used to studying situations that do not fit the established system of disciplines that separates social issues from natural scientific ones, economic questions from political ones, and human activities from nonhuman factors. To analyse what went on in a laboratory in California, in Pasteur's *Mémoire*, and in late nineteenth-century developments in French agriculture and public health, or to answer the question who or what had killed *Aramis*, time and again the trick was to study the translations by which human and nonhuman actants form actor-networks that make up established facts, an ill-fated innovative public transport system, or turned Pasteur into a celebrity. So we might expect that the same trick will probably do also for understanding a news story in *Le Monde*.

In *We Have Never Been Modern*, Latour aims higher than merely repeating this message. Like Weber, he wants to understand what made the West exceptional and successful. And – going beyond Weber – he wants to understand how this success gave way to a world that today has to deal with an abundance of issues in which science and politics, technologies and ethics seem to have become tied together in Gordian knots. To address these concerns, Latour argues, another dimension of analysis has to be added. Science and technology studies have brought only half of the story. The leading hypothesis of *We Have Never Been Modern* is that

the word 'modern' designates two sets of entirely different practices which must remain distinct if they are to remain effective, but have recently begun to be confused. The first set of practices, by 'translation', creates mixtures between entirely new types of beings, hybrids of nature and culture. The second, by 'purification', creates two entirely distinct ontological zones: that of human beings on the one hand; that of nonhumans on the other. Without the first set, the practices of purification would be fruitless or pointless. Without the second, the work of translation would be slowed down, limited, or even ruled out. The first set corresponds to what [Latour has called] networks; the second to what [he] call[s] the modern critical stance. The first for example, would link in one continuous chain the chemistry of the upper atmosphere, scientific and industrial strategies, the preoccupations of heads of state, the anxieties of ecologists; the second would establish a partition between a natural world that has always been there, a society with predictable and stable interests and stakes, and a discourse that is independent of both reference and society. (NBM: 10–11)

So we have a triple task: we have to study the practice of translation, the practice of purification and the relation between the two. The first task has been taken up by science and technology studies. Weber had already started the second one. Purification, that is, distinguishing between on the one hand the sphere of human values and interests and on the other the disenchanted sphere of the inanimate world, is closely related to what Weber called rationalization. However, there are two crucial differences. Latour claims that purification is a practice, not an acquired attitude; purification requires *work*. And with regard to the relation between these two practices, he claims that the practice of purification comes second to the practice of translation.

If purification comes second to translation, one should be able to study its emergence out of translation work. In a long discussion of Shapin and Schaffer's (1985) brilliant study *Leviathan and the Air-Pump*, Latour argues that the authors had done just that (Latour 1990a; NBM: Chapter 2). Explicitly intended as an exercise in the sociology of scientific knowledge (Shapin and Schaffer 1985: 15), *Leviathan and the Air-Pump* reconstructs and analyses a debate between Boyle and Hobbes in the 1660s and early 1670s.

Boyle and Hobbes agreed on almost everything. They both "want a king, a parliament, a docile and unified church, and both were fervent subscribers to mechanistic philosophy" (NBM: 17). Their dispute concerned the legitimacy of knowledge produced on the basis of experiments. Boyle claimed that his experiments, overseen by reliable gentleman-witnesses, produced authenticated facts. Hobbes demanded the certainty of mathematical proof, the kind of knowledge that can be checked not only by a few gentleman friends gathered in a private space, but publicly, step by step, by anybody. But behind the epistemological issues, big political questions about social order loomed for which Boyle and Hobbes offered rivalling solutions.

One solution (Boyle's) was to set the house of natural philosophy in order by remedying its divisions and by withdrawing it from contentious links with civic philosophy. Thus repaired, the community of natural philosophers could establish its legitimacy in Restoration culture more effective to guaranteeing order and right religion in society. Another solution (Hobbes's) demanded that order was only ensured by erecting a demonstrative philosophy that allowed no boundaries between the natural, the human, and the social, and which allowed for no dissent within it. (Shapin and Schaffer 1985: 21)

To explain why Boyle's solution prevailed, Shapin and Schaffer proceed in three steps. Displaying in fascinating detail the conventions and the crafts the two opponents required for producing authenticated knowledge, and by laying out the new style of reporting and discussing facts Boyle introduced and the proliferation of scientific instruments, Shapin and Schaffer argue that the solution to the (epistemological) problem of knowledge involved laying down the rules and conventions of the intellectual community, with Boyle and Hobbes opting each for a different form of intellectual life. Secondly, Shapin and Schaffer argue that the knowledge thus produced became an element in political action in the wider society. Finally they claim that "the contest among alternative forms of life and their characteristic forms of intellectual product depends upon the political success of the various candidates in insinuating themselves into the activities of other institutions and other interest groups. He who has the most, and the most powerful, allies wins" (Shapin and Schaffer 1985: 342).

As one of the founders of the Royal Society and throughout his life being one of its most notable fellows, Boyle had the better hand. He could mobilize powerful friends. Experimental science was instituted as a source of authenticated knowledge, separated from politics, state and religion. Already in 1663, drafting the constitution of the Royal Society, Hooke articulated the divide between, on the one hand, the business of science and, on the other hand, politics, religion and morality:

The Business and Design of the Royal Society is – To improve the knowledge of natural things, and all useful Arts, Manufactures, Mechanick practices, Engynes and Inventions by Experiments – (not meddling with Divinity, Metaphysics, Moralls, Politicks, Grammar, Rhetorick, or Logick). (cited in Ornstein 1963 [1913]: 108 n. 63)

Boyle won. However, in the last sentences of their book, Shapin and Schaffer claim: "[a]s we come to recognize the conventional and artefactual status of our forms of knowing, we put ourselves in a position to realize that it is ourselves and not reality that is responsible for what we know. Knowledge, as much as the state, is the product of human actions. [So, Boyle may have won the dispute, but] Hobbes was right" (Shapin and Schaffer 1985: 344).

Given his former debates with Schaffer and other sociologists of scientific knowledge, we cannot expect Latour flatly to accept this conclusion. Shapin and Schaffer's claim to have explained the authentication of the content of experimental knowledge by political action in the context of British society fails according to Latour, because "neither existed in this new way before Boyle and Hobbes reached their respective goals and settled their differences" (NBM: 16).

So, much to the chagrin of the authors, Latour rewrote their argument, cleverly turning *Leviathan and the Air-Pump* into another kind of study, namely a comparative anthropological investigation of the dual process of establishing science and politics as separate domains, rather than an exercise in the sociology of scientific knowledge. On Latour's reading, what made Boyle prevail and Hobbes lose the dispute was something else than Boyle having powerful gentleman allies.

Boyle had acquired an airpump (invented by Von Guericke in the 1650s) and he claimed as a matter of fact that with this machine he had created a vacuum in a glass globe (the airpump's receiver). This claim was unacceptable for Hobbes, who adhered to 'plenism', a long standing philosophical doctrine to which no genuine potentiality of being can remain unfulfilled and that, as a consequence, conceived a vacuum to be non-existent and even unthinkable. However, as Shapin and Schaffer had noted, Hobbes's opposition was not only based on philosophical grounds. He was driven by fear of disruption of social order, the same fear that had motivated him to write his famous treatise on political philosophy, *Leviathan* (Hobbes 1980 [1651]). Religious controversies about esoteric issues had already resulted in bloody civil wars. If a few gentlemen could claim to have artificially produced something non-existent and even unthinkable, a vacuum, in their private rooms, soon new quarrels would start and before long civil wars would break out again.

Therefore Hobbes took the trouble to try to refute Boyle's claim. Based on his plenist's principles, he argued that when Boyle had exhausted the air from the glass globe, something else, a substance he called aether that surrounds the Earth and which he supposed to be so subtle that it could penetrate anything, including glass, must have filled the void. To reply to this objection, Boyle made his experiment more sophisticated (NBM: 22; Shapin and Schaffer 1985: 181ff). He put a little feather in the glass globe. If the receiver, exhausted from common air, would be filled with aether, jets of aether caused by the Earth (and thus the receiver) moving through the aether, would certainly move the feather. But the feather didn't move. All witnessing gentlemen agreed on that. So, Boyle concluded, his experiment proved Hobbes's speculative reasoning to be wrong.

So, Latour concludes, it was not an old-boys' network of powerful friends, but a hybrid alliance of an airpump, a few gentlemen, and a little feather that made Boyle's case. Of course, this event alone cannot *explain* the establishment of experimental science as a source of authenticated knowledge, separate from politics, religion and morals. But 'explanation' is not what Latour is after. He *redescribes* the crucial event in the Boyle–Hobbes debate. In Latour's description, Boyle's innovative step was to apply the old repertoire of penal law and biblical exegesis to a new point. Earlier, witnesses had always been human or divine. In Boyle's texts, a new kind of witness is recognized, "inert bodies, incapable of will and bias but capable of showing, signing, writing and scribbling on laboratory instruments before trustworthy witnesses" (NBM: 23). Boyle introduced "the testimony of nonhumans" – in the above experiment, the testimony of a little feather – a kind of witnessing even more trustworthy than humans. But in the court of science, this testimony will lose its power when other considerations – e.g. political, religious or moral ones – get the upper hand. So, to allow nonhumans to provide testimony, all connections with political and religious issues had to be broken off. To allow experimental science to produce authenticated knowledge about matters of fact, a two-houses solution was invented: in one house, matters of natural philosophy, that is, science, would be decided; in the other house, all other issues. "Not meddling with Divinity, Metaphysics, Moralls, Politicks, Grammar, Rhetorick, or Logick" in the house of science was an offer the experimental scientists were more than willing to make.

The dice had been cast. Powers, competences and responsibilities were distributed. A divide had been instituted. By introducing the testimony of nonhumans, Boyle and his followers had found a way to scientifically represent facts of nature – objects. In *Leviathan*, by creating the citizen who in exchange for safety and peace cedes rights to the artificial construction of the Sovereign, Hobbes had invented the modern political and social subject. Simultaneously, a differentiation of (scientific) content and (social and political) context had been created. This, Latour claims, is what Shapin and Schaffer's *Leviathan and the Air-Pump* documents – provided that their book is read as a study of the simultaneous birth of both modern science and modern politics, each with their own vocabularies, procedures, institutions and vocations. Read in the right way – that is to say: not as an exercise in sociology of scientific knowledge but as a comparative, anthropological one – *Leviathan and the Air-Pump* shows that the divide between science and politics was created out of the translation work performed by Boyle (on the object-side) and Hobbes (on the subject-side). Once the divide was in place, "the word 'representation' [would] take on two different meanings, according to whether elected [human] agents or [nonhuman] things are at stake" (NBM: 29). From now on, human values and interests were to be represented in politics, and facts of nature by science. Each had got its own institutions, procedures, rituals and meeting rooms.

5.2   The modern Constitution

How did the distribution of competences and responsibilities that had been reached in a small circle in the seventeenth century spread out to become virtually unquestioned in the centuries to come? That part of the history of the West has still to be written. To provide an answer to the question, Latour shifts registers. Leaving ethnography for the moment, he engages in philosophical analysis and speculation. He presents the arrangement that came out of the debate between Boyle and Hobbes as being laid down in a 'modern Constitution' that separates the powers of politics and science, to argue that the arrangement remained in place for centuries because the internal, intellectual structure of this Constitution shields it from criticism and change.

Political constitutions divide the various branches of government. They are the outcome of long political discussions before being ordained and established. Typically, they contain procedural guarantees to prevent the casual or frivolous overthrow of the established constitutional arrangements, for example by demanding that amendments to the constitution require new elections and a two-thirds majority in both chambers of parliament. The Constitution Latour writes about (Constitution-capital-c to distinguish this one from political constitutions) does not separate branches of government, but the powers of, on the one hand, 'Nature' – the common denominator for objects and of brute natural facts, the domain of the natural sciences – and, on the other hand, 'Society' – comprising human interpretation and activity, values and symbolic constructions, the domain of politics and culture. This Constitution didn't get shaped in political discussions. It evolved out of the mainstream of seventeenth, eighteenth and nineteenth-century philosophy and was anchored in the division between the institutions of science and of politics.

What Latour calls 'the modern Constitution' is an aggregate that introduces a host of – partly – overlapping dichotomies that contrast mind and matter; humans and nonhumans; values and facts; the Subject of knowledge and the known Object; human society, culture and politics on the one hand, and nature and science on the other hand. One would search in vain for a single philosopher who defends all of these dichotomies together in raw form; Latour's 'modern Constitution' is a blend of Descartes' ontological distinction between the realm of thinking (*res cogitans*) and the realm of things (*res extensa*), and Kant's critical philosophy that argued that the world we know is the phenomenal world, the world of our experiences mediated by the pure forms of intuition and the categories of human understanding, and not the (noumenal) world of things-in-themselves. Rather than taking the whole of the modern Constitution on board, after Kant philosophers took it as the *starting* point for their reflections (cf. e.g. Schnädelbach 1983). With Kant, they took for granted that Newton had shown that science can inform us about the natural world, to subsequently shift their attention to the Subject-side of the divide, to mind, culture, society and language, to the foundations of the *Geisteswissenschaften* and to the question of what remained to be said about religion and ethics once reason had entered "the secure path of a science" (Kant 1956 [1787]: Bix). Their attempts to bridge the gap between *Sein und Denken*, between the Object- and the Subject-pole, started from the Subject-pole. Idealism reigned. Even philosophers who usually are not labelled as 'idealists' (e.g. the early Wittgenstein, cf. Stenius 1964) followed the strategy of deducing whatever they had to say about the ontological structure of the world from their reflections on knowledge, language and mind.

True, not all philosophers were convinced and the nineteenth-century discovery that not only do societies and cultures evolve but that also geological formations and animal species had changed over time presented quite a puzzle (Mandelbaum 1971). 'Life' became a heated subject of debate in nineteenth-century philosophy; vitalists argued that living creatures have their own ontology, separated from inanimate nature. Schopenhauer (1977 [1818]) boldly stated that to deny nature to be nothing more than appearance would be a form of solipsism that belongs only to the madhouse. In his philosophy, not only humans but also inanimate objects show a 'Will'. Dissociating 'will' from its psychological connotations, straining language, Nietzsche even argued that inanimate beings *are* (not: *have*) a 'will' of their own (cf. Nehamas 1985: ch. 3). But the mainstream of the European tradition placed the Object-side, Nature, beyond philosophical analysis. To be informed about Nature, one should consult a scientist, not a philosopher. In spite of generations of philosophers arguing for more subtle distinctions, the modern Constitution – the fundamental divide between the poles of Nature and of Society – was firmly in place.

Of course, the philosophers didn't deny that there is a world out-there to which humans intimately connect and that shapes human relations: people need food, water, energy, they share a house, operate machines together, exchange material items and use a wide range of technologies to communicate and connect with each other. If there were not a material world, society would not last a minute and without their bodies being fed, human minds would soon have other concerns than theoretical philosophy.

So why did the modern Constitution remain firmly in place, in spite of generations of philosophers arguing for more subtle conceptual distinctions and in spite of the mundane considerations that speak to the contrary? On Latour's speculative analysis, like a political constitution, the modern Constitution, too, came with guarantees that prevented overthrowing its arrangements lightheartedly. The internal architecture of the modern Constitution led us into an intellectual game the outcome of which is decided: 'Heads I win, tails you lose.'

These are the rules of this game, the "guarantees of the modern Constitution" (NBM 29–35). *Guarantee (1)*: The Constitution declares that (*ontologically*) Nature is not our construction (i.e. Nature is transcendent), but Society is (i.e. Society is immanent). Nevertheless, with concern to *knowledge*, the tables are turned. *Guarantee (2)*: The Nature science talks about is the phenomenal world, mediated by the categories of human understanding and often artificially constructed in a laboratory (Nature is immanent); whereas Society is given as an ensemble of social facts that social scientists may investigate (Society is transcendent). So the Constitution paradoxically declares: even though we construct Nature, Nature is *as if* we did not construct it. A similar paradox applies at the other side: though we do construct Society, Society is *as if* we do not construct it. Both paradoxes are evaded by *Guarantee (3)*: There shall exist a complete separation between the natural world and the social world; and there shall exist a total separation between the work of translation (that goes on in constructing nature in laboratories and society in social practices) and the work of purification (that goes on in stating that the two spheres are ontologically separate). Finally, Latour lists *Guarantee (4)*: God is removed forever from the dual social and natural construction. But, Latour observes, this guarantee is offered only half-heartedly. God remains presentable and usable. In case of conflict between the laws of Nature and those of Society, the right to appeal to the transcendence of God is reserved. As an effect, spirituality was reinvented: "[m]odern men and women could [. . .] be atheists while remaining religious" (NBM: 33) – although perhaps only in their heart of hearts.

With these four guarantees, the modern Constitution presents simultaneously a set of anchor points and a set of possibilities that provide an unbeatable set of four contradictory resources for critique (NBM: 36). Nature's laws are given, but we have unlimited possibilities to experiment and to develop technologies; men are free, but social science may establish the laws of human behaviour and stable patterns of social action we can do nothing against. So Moderns "can mobilize Nature at the heart of social relationships, even as they leave Nature infinitely remote from human beings; they are free to make and unmake society, even as they render its [social, economic and psychological] laws ineluctable, necessary and absolute" (NBM: 37). 'Heads I win, tails you lose.'

However, in spite of their dualistic worldview, the Moderns continued to mix humans and nonhumans "under the table", in their translation work. They created an abundance of new *hybrids*. Laboratories kept producing knowledge that could be used for developing technologies that increasingly began to make up the fabric of modern society – vaccines, technologies for long distance communication, plastics, computers, genetically modified organisms, etc. However, in the ontology of the modern Constitution there is no place for these hybrids. So they were purified. The results of the translation work in scientific laboratories were taken for 'discoveries' of brute facts of Nature, attained by scientists with rational minds and methods. Also the technologies that came out of laboratories were purified. If their origin was discussed, they were conceived as the products of value-free science; but their use and social impact were conceived as a matter that might be discussed in politics and studied by social scientists. After this clean up, no further discussion was deemed necessary. Amazingly, the role of technology in making up modern societies and in establishing social order was not studied. As to philosophy of technology: it never took off.

The Constitution's Guarantee (3) had made the practice of translation invisible, under the radar of the official view. But this didn't limit the practice of translation in any way. It continued at great pace. Without it,

> [. . .] the modern world would immediately cease to function. Like all other collectives [the term Latour uses to designate associations of humans and nonhumans] it lives on that blending. On the contrary (and here the beauty of the mechanism comes to light), *the modern Constitution allows the expanded proliferation of the hybrids whose existence, whose very possibility, it denies.* (NBM: 34)

The Constitution had another important effect. Apart from the divide between Nature and Society, it also created a divide between Us, Moderns, and Them, premodern peoples, who continue to confuse the domains of humans (Society) and nonhumans (Nature) that the Constitution declared to be separated. Placing events either before or after the advent of the Constitution, the Constitution created its own time-line with a clear break. Before: traditionalism, confusing things and men; after: modernity and the future, separating the disenchanted world of brute facts from the domain of the mind, values and meanings. The internal divide between Nature and Society seamlessly led to an external divide between Us (Moderns) and Them (premoderns).

Weber's analysis of the exceptionality of modern Occidental culture and society is a straight consequence of identifying modernity in the terms of its Constitution. Weber was right, that is to say: once we accept the terms of the modern Constitution. No, Weber was wrong: we have never been modern; next to the practice of purification the practice of translation has also continued. And this practice led to new concerns, partly reformulating the two downsides of modernity Weber had already observed – loss of freedom and loss of meaning – but also introducing new ones: the scientification of politics, the ideology of industrial society, efficiency, and 'one dimensional man' (Marcuse 1964). What Weber had singled out as the Occident's exceptional form of rationality, *Zweckrationalität*, came to be perceived as a 'halved rationality' (Habermas 1969). Where could a broader concept of rationality be found? In search of it, the Frankfurt School and other critical theorists again focused exclusively on the Subject-pole, fortifying the modern Constitution further.

Latour displays little patience with critical theory. He takes a more radical turn. He sets out to introduce a conceptual framework that will let us see what the Moderns, looking through the lens of the modern Constitution, failed to see. To the practice of purification, he adds the practice of translation, the practice from which hybrid beings emerge, to give it pride of place. The ontology developed in almost twenty years of science studies was to replace the modern Constitution. Kant had revolutionized philosophy by turning Descartes' philosophy inside-out; Latour's philosophy drains the modern Constitution by prioritizing the middle ground, the production of hybrids, rather than the two poles of Nature and Society. He replaced the modern Constitution by what has been called "hybrid thoughts in a hybrid world" (Blok and Jensen 2011).

5.3 Relationism

Hybrid thoughts? For a philosopher, that is not necessarily a compliment. So let's first describe in broad strokes what Latour suggests as an alternative to the modern Constitution, to subsequently get to the details.

What are the 'hybrids' about which Latour talks? The term itself shows embarrassment. A 'hybrid' denotes something made by conceiving two different elements, a mixture – but a mixture of what? Humans and nonhumans? Nature and Society? Construction and reality? Subject and Object? But then we define 'hybrids' in terms of the language of the modern Constitution – the conceptual architecture that, Latour suggests, needs to be replaced by something else.

Again, Latour runs against the problem anyone encounters who wants to formulate a radical new idea: he has to communicate his new thoughts in the old language. In his work in science and technology studies, Latour could take his readers by the hand. The ethnographic material he brought to the fore allowed him to point his readers to what they had failed to see before; meanwhile he could introduce, step-by-step, a new terminology for describing what he wanted his readers to notice. In *We Have Never Been Modern* – and in his later work *An Inquiry into Modes of Existence* (cf. Chapter 6) – Latour presents the conclusions of his earlier work in science and technology studies in general terms. He wants his modern readers to get rid of the conceptual framework that comes naturally to them and that is officially institutionalized in the separation of the representation of humans by politics and of nonhumans by science; he wants them to abandon the conceptual distinctions that evolved out of the attempts of eighteenth and nineteenth-century philosophers to account for science in episte- mological terms, that is, in terms of the relation between an active knowing Subject and inert Objects lying waiting for centuries to be 'discovered' by a scientific genius. He asks them to replace the *Weltanschaung* that comes naturally to them with a philosophy that prioritizes *existence* over essence (i.e. eternally given – but hidden – substance) and that gives *ontology* pride of place over epistemol- ogy. He wants them to turn their *Weltanschaung* inside-out, to define hybrids not in terms of the constituents the modern Con- stitution takes for being given, but rather to give them pride of place and to see what then appears as given, as the *outcome* of processes, of translations.

But how to argue for this transition without using the old terms? Latour asks his readers to acknowledge that we are uncertain about what *is* action, a social group, a fact, technology, what *is* law, science and politics and to abandon not only what they thought they knew about knowledge, but also what they thought they knew about the world. To make this move, Latour invites his readers to have a closer look at who we are and what we practice.

As we have seen above, the Constitution's internal divide between Nature and Society led to an external divide between Them – premodern peoples – and Us – the Moderns. The Moderns locate themselves on a timeline that runs from past to present – and further into the future – with a radical break in the past, the birth of modernity; a break that took place somewhere in the sixteenth and seventeenth centuries, when – as Weber suggested – Platonism married Renaissance experimentalism to change Man's conception of the universe and his own place in it, while simultaneously modern politics was invented by Hobbes.

However, as Latour points out, our actual practices are more heterogeneous.

> I may use an electric drill, but I also use a hammer. The former is thirty-five years old, the latter hundreds of thousands. Will you see me as a DIY expert 'of contrasts' because I mix up gestures from different times? Would I be an ethnographic curiosity? On the contrary: show me an activity that is homogeneous from the point of view of the modern time. Some of my genes are 500 million years old, others 3 million, others 100.000 years, and my habits range in age from a few days to several thousand years. As Péguy's Clio said, and as Michel Serres repeats, 'we are exchangers and brewers of time' (Serres and Latour 1995). It is this exchange that defines us, not the calendar or the flow that the [M]oderns had constructed for us. (NBM: 75)

Brewers of time? So are we still confused, living for one bit in modernity and for another bit in traditionalism? No. We need a more refined conception of time. The whole idea of a stable 'tradition' is a misconception. Most 'traditions' are modern inventions; for example, the traditional Scottish kilt was invented at the beginning of the nineteenth century (Trevor-Roper 1983). "One is not born traditional; one chooses to become traditional by constant innovation" (NBM: 76).

Nevertheless we actively sort out elements belonging to different times. But to sort events in time, one has to first introduce calendars and clocks. *"It is the sorting that makes the times, not the times that make the sorting"* (NBM: 76). Calendar time orders events on a chosen timeline. Crucial events are given a date. 1812: Battle of Borodino. 1858: Pasteur discovers 'lactic yeast', the birth of microbiology. 1995: the first *United Nations Climate Change Conference*. We may also put dates to the emergence of new geological formations or to the evolution of a species on our calendar, e.g. to record that the geological epoch of the Holocene started 11,700 years before the present.

But calendar time is not 'history'. As we have seen before (ch. 3), for Tolstoy 'history' does not denote a series of events in calendar time, but *infringements* of life. Latour cites a similar idea from Péguy's *Clio*. Comparing Victor Hugo's "terrible" play *Les*

Burgraves with a little phrase from Beaumarchais, Péguy (cited in NBM: 68) wrote:

> When I am told that Hatto, the son of Magnus, the Marquis of Verona, the Burgrave of Nollig, is the father of Gorlois, son of Hatto (bastard), Burgrave of Sareck, I learn nothing,' she [Clio, the muse of history] says. 'I do not know them. I shall never know them. But when I am told that Cherubino is dead, *in a swift storming of a fort to which he had not been assigned*, oh, then I really learn something. And I know quite well what I am being told. A secret trembling alerts me to the fact that I have heard.'

A 'historical' event is an event that *happens* to someone, an event that affects one's *existence*. 'History' situates events with respect to their intensity. For Tolstoy 'history' is limited to the human domain. It denotes an event that happened to someone's, a group's, or a nation's existence. Latour suggests that the notion of history also applies to nonhumans, to note that something has happened to it – that is, has affected its *existence*. And he claims that it not only applies to nonhumans that – as we know since the nineteenth century – have evolved over time (like biological species and geological formations), but to *any* nonhuman and (when we put calendar dates to the events that happen to them) on scales that are much smaller than geological or evolutionary time. For example, when a nonhuman is put on trial in an experiment, this event may become part of the history of that nonhuman.

To get the idea, reconsider the discovery of lactic yeast by Pasteur. What *is* the object Pasteur discovered in his experiments? As we have seen before (§§ 2.3 and 2.4), it started as a barely visible grey mass in ordinary lactic fermentation, was turned into something Pasteur could collect, transport and sprinkle, was subsequently identified by Pasteur as a living organism – like brewer's yeast but different – to be turned in experiments of later generations of microbiologists into 'lactic acid bacteria' (plural), that is, not one specific micro-organism but a whole clade of bacteria.

The modern Constitution suggests that in the course of a century of research humans have discovered the true nature of the microbes that cause souring of milk, their essence. By now, we know that what Pasteur called 'lactic yeast' is, in fact, a whole clade of microbes, rather than a specific member. That, however, is the epistemological interpretation – that is, the story of the way in which in the course of (calendar) time the Object appeared to the knowing Subject.

Latour suggests that to account for what has happened we better tell this story in ontological terms. In that story, events happened that changed this being's *existence*. It had quite some career. It started as a ferment of ordinary lactic fermentation (a dubious – chemical or biological? – entity that even might not exist at all); it became the barely visible grey mass that drew Pasteur's attention and that eventually led him to claim it to be a kind of yeast; to become a hundred years later a family of bacteria. What *happened* to this being? In 1857, it met Pasteur, who provided a culture in which it could multiply at greater speed then ever before. As a consequence, it acquired a new characteristic: it became a clearly visible being. Some decades later, in the experiments of microbiologists of a later generation, it changed its existence again to become a whole family of bacteria. That is the *history* of this being.

Sartre (1966 [1945]: 32–33) famously wrote "*l'existence précède l'essence*". Man first of all exists, encounters himself, surges up in the world – and defines himself afterwards. Latour extends the phrase to apply also to nonhumans: their existence precedes their essence too. In the course of its history, the existence of a nonhuman may become more stable, at which point we may guess that we have identified its essence (while acknowledging that future research may correct or even overthrow this guess – yes, we have read Popper). But in ontological terms, nonhumans also have a history; they are affected by what happens to them. They are not the Object of the modern Constitution, an essence waiting to be discovered; they are beings whose *existence* changes in the course of their history. They have a 'variable ontology' – their characteristics as an existent change when something happens to them.

Of course, we may subsequently put some dates next to the events that happened to a being, to locate these events in calendar time. So we may speak of the being that in 1858 was identified as 'lactic yeast' and about the being that more than a century later is identified as 'lactic acid bacteria'. However, to carelessly mix up calendar time with history, thinking that although time has flown it has remained the same being with a stable substance (that is, the Object the modern Constitution), is asking for conceptual trouble. No, Pasteur's lactic yeast and today's lactic acid bacteria have different ontologies. In ontological terms, they have quite different existences; they are different beings.

Moving back in time, this leads to a conclusion that sounds completely weird to modern ears: 'lactic yeast' did not exist before Pasteur. Of course, before Pasteur, milk, when left alone for a few days, went sour. But to say that in the old days either 'lactic yeast' or 'lactic acid bacteria' caused the souring of milk is an anachronism. Whatever caused milk to turn sour before Pasteur *became* 'lactic yeast' – a being with new characteristics, namely a being that is visible to the human eye and that can be isolated, sprinkled and transported – only after having been translated in Pasteur's experiments (PH: ch. 5; Latour 1996c; 1999c; 2008).

Serres suggested the concept Latour uses to refer to beings with a variable ontology; he calls them 'quasi-objects'. It is a concept that may easily confuse us. Quasi-objects are not substances plus or minus something. Quasi-objects are entities that we perceive (using the modern Constitution's language to express what we perceive) as objects, as given things, but that have an entirely different ontology from the Objects of the modern Constitution. They are not defined in terms of a (hidden) substance that in the course of the development of scientific knowledge may be 'discovered'. They exist; what they are depends on what has happened to them, on the various translations that they have become involved in. What caused milk to get sour before Pasteur only became lactic yeast in Pasteur's experiments, to change again later from a specific microorganism to a whole family of microbes. It didn't just change its name; a lot happened to it; in the course of its career it became another *being* – a being with new characteristics, a being caught up in and defined by new networks.

The concept of 'quasi-object' bends the language of the modern Constitution for the purpose of understanding the ontology Latour is after. It may help understanding what Latour is after, but may also cause confusion. Isn't there a better term? In the 1990s, Latour suggests 'proposition', a concept, drawn from Whitehead, that does not designate a statement that may be true or false, but any actant (human, nonhuman, or any hybrid association of humans and nonhumans) that proposes itself to other actants to become translated into another (joint) being (PN: 247; PH: 309; the definitions differ slightly). But this term, too, didn't last long. It took Latour twenty-five years before he found in *An Inquiry into Modes of Existence* the words to express the difference: they are beings-as-other, rather than (the modern Constitution's) beings-as-being. Terrible jargon, indeed. We'll get to it in §6.1.

Having introduced 'quasi-objects' to replace the Object of the modern Constitution, we may expect Latour also to replace the Subject of the modern Constitution by 'quasi-subjects'. In this case, we may seem to have less trouble in understanding what is meant.

Nineteenth-century philosophy already had argued that the Subject has history; novels and Freud's psychology taught us to conceive persons in terms of their biographical past, in terms of what has happened to them as a child and in their personal relations. However, Latour's reasons for speaking about 'quasi-subjects' diverges from these traditions. The main reason is not to allow societies, cultures and human minds to have a history, but to allow nonhumans to be part of the existence of (quasi-) subjects. In *Reassembling the Social*, we have already become acquainted with the idea that actors, human beings, persons, are not stand-alone entities. They too have variable ontologies. For their existence, to act, they require plug-ins, scripts, and attachments. Their action may be overtaken; their instruments may make them do something. To speak about the existence of societies, cultures and the human mind, we have to take also the nonhumans into account. They happen also to human beings. Yes, Pasteur happened to the microbes, but the microbes also happened to Pasteur, to make him the person France would celebrate as its genius. We cannot understand who humans are, without taking the nonhumans, the quasi-objects, into account.

In a sense, by distinguishing between time and history, Latour has transferred to ontology a key idea from Einstein's theory of relativity. Time is not a given framework, Einstein taught physicists; to sort events in time, one needs calendars and clocks. For Latour, existence is relative; to register how (human and nonhuman) beings exist, one has to focus on their relations, mediated by instruments, experiments, by other humans and nonhumans, whatever, to detect what has happened to them, in their history.

So is Latour a 'relativist' – a reproach he has had to defend himself against time and again (PH: 4–23)? To avoid the confusions the term 'relativism' invokes in philosophy (where it is mainly associated with moral or epistemological relativism, poor man's philosophies that Latour never held), we better avoid this label. Better to call his ontological position *relationism* (NBM: 114). So to call Latour a philosopher of 'hybrid thoughts' is a bit unfair. He defends a systematic, coherent philosophical position, namely a *relationist* ontology, a philosophy that gives pride of place to relations, to translations, rather than to what enters into these relations.

It has been quite a philosophical *tour de force* indeed. But remember that the reasons for engaging in it were concrete questions. For a long time, we could live with the modern Constitution to benefit from its fruits. It allowed the proliferation of ever more hybrids, many of which helped to advance health and prosperity. But by approaching the end of the twentieth century, concerns about technologies, ecological problems, and advances in biotechnology and medical science that elicit ethical questions have begun to dominate the political agenda. They put pressure on the established way of understanding the relations between nature, science and politics. They signal that we need a better way of understanding what we have been doing, who we are and how to live in a world that includes both humans and nonhumans.

This is what urged Latour to dive deep into philosophical waters. "We want the meticulous sorting of quasi-objects to become possible – no longer unofficially and under the table, but officially and in broad daylight" (NBM: 143). Indeed, in this desire to bring to light, to make public, Latour adheres to the intuitions of the Enlightenment. But to get quasi-objects in public view, we have to abandon the modern Constitution that emerged out of the Enlightenment. We need a philosophy that allows detecting variable ontologies: a relationist, empirical philosophy.

5.4   Cosmopolitics

In *We Have Never Been Modern*, Latour advances the philosophy that he deployed – for a long time implicitly – in his work in science and technology studies to understand the political issues that the world faces today, in particular in relation to ecological problems.

New political thinking is called for. On that Beck and Latour agree. The political philosophy that came out of the modern tradition did not anticipate that governments would have to concern themselves with administrating the global climate, the ozone layer and environmental problems that transgress national borders. Conceiving politics to be exclusively the realm of human values, interests and ambitions, it focused on analysing the power-relations between citizens and the state. But 'green' political philosophies that have inspired ecological movements do not fare better, Latour argues. Calling on Society to protect Nature, they only seek "to position themselves on the political chessboard without redrawing its squares, without redefining the rules of the game, without redesigning the pawns" (PN: 5). They appeal to Nature as an authority, only to find scientists fiercely debating what Nature tells us. To develop a more realistic political ecology, the concept of Nature has to go; we have to reconsider what "nature, science, and politics have to do with one another" (PN: 6).

In the final sections of *We Have Never Been Modern*, Latour already outlined his political philosophy. It ends with what he calls a 'Parliament of Things' in which both humans and nonhumans are properly represented – a rather troubling concept: how are the speechless nonhumans to be represented; and if we allow scientists to do so, does Latour propose that scientists become members of parliament, like the military taking up guaranteed seats in parliament in some dictatorially governed countries? Our confusion will hardly be eased by Latour's statement that "we do not have to create this Parliament out of whole cloth [. . .], [w]e [only] have to ratify what we have always done [. . .]" (NBM: 144).

Fortunately, in *Politics of Nature*, he offers more detail. He defends a 'political ecology' that "has nothing at all to do with 'nature' – that blend of Greek politics, French Cartesianism and American parks" (PN: 4–5) – a political philosophy that designates "the right to compose a common world [of humans and nonhumans], the kind of world the Greeks called a *cosmos*" (PN: 8), a philosophy for *cosmopolitics*.

Politics of Nature is dedicated to among others Isabelle Stengers, who introduced the concept of 'cosmopolitics' (Stengers 1996–1997). The term is rather grandiose and needs some unpacking.

In the first place, 'cosmopolitics' is not to be taken for 'cosmopolitanism', the idea – defended in various versions by, among others, the Stoics and Kant – that all human beings are (or can and should be) citizens of a single community. Secondly, in contrast to what the term 'cosmos' might suggest, we do not have to look at the starry heavens and to think about the universe as a whole. We should not look up, but down. The 'cosmos' Latour's philosophy addresses is an assembly of mundane assemblies; it refers to the one Earth that humans share with other humans and with nonhuman entities; the concept acknowledges – with Pasteur – that there are more of us than we thought and that there is a multiplicity of interests that have to be taken into political account – human ones as well as the interests of nonhumans.

This leads to the second component of the term 'cosmopolitics', 'politics'. Given the multiplicity of beings involved, to institute some order and peace, recourse to an encompassing entity or idea – e.g. Reason, Nature, or God – or some arbiter agreed-upon by all is not available. That, according to Latour, is what 'politics' is about: instituting order in the absence of any a priori point of agreement. With some reluctance, Latour cites Schmitt (1996 [1932]), a "toxic and nevertheless indispensable" (FG: 295) author, who claimed that "[t]he specific political distinction to which political actions and motives can be reduced is that between friend and enemy" (Schmitt 1996 [1932]: 26). Schmitt's further explanation may not necessarily comfort us:

> The political enemy need not be morally evil or aesthetically ugly; he need not appear as an economic competitor, and it may even be advantageous to engage with him in business transactions. But he is, nevertheless, the other, the stranger; and it suffices for his nature that he is, in a specifically intense way, existentially something different and alien, so that in the extreme case conflicts with him are possible. These can be neither decided by a previously determined general norm nor by the judgment of a disinterested and therefore neutral third party. (Schmitt 1996 [1932]: 27)

In spite of Schmitt's bellicose language, this sets the stage for what is the lead-question of Latour's cosmopolitics: how to establish a common, yet plural world, one in which we are confronted and challenged by – human and nonhuman – others, without taking recourse to some pre-given, agreed-upon authority?

For the sake of clarity, the argument of *Politics of Nature* may be divided into four steps.

In the first step, Latour asks us, once again, to give ontology rather than epistemology pride of place. In *Politics of Nature*, he does so by urging us "to get out of Plato's cave" (PN: 10ff). Why does he step so far back in time? He tries to kill two birds with one stone. Plato's allegory of the cave suggested a division between on the one hand the Ideas of the True, the Good and the Beautiful and on the other the world of the *hoi polloi*, the common people, who are chained for life in a cave and who, taking appearance for reality, chatter about the shadows that are thrown on the wall they face. Only those few who can break away from their fetters and leave the cave will be able to see the True, the Good and the Beautiful. Bringing the news about what they have found outside back to the cave, they can start to teach and lead the *hoi polloi*. A powerful – and as it turned out often brutal – motive was seeded into the cultural, political and religious traditions of the West (ICON): to live a truly good life requires first and foremost the destruction of false images; it takes people who have seen the light to break the spell. Not only philosophers embraced that message. It motivated centuries of religious wars and ideological disputes.

In *Politics of Nature*, Latour merges Plato's philosopher with the modern scientist. That, of course, is incorrect. Modern scientists only claim to have exclusive access to nature, to Truth, not to the Good and the Beautiful. In terms of Plato's allegory, the scientific revolution and the philosophical tradition that accepted the role of the new science projected the Good and the Beautiful back into the cave, to become a matter of value-disputes, of conflicting subjective opinions.

However, two important assumptions were retained: the idea that public life is organized into two houses – which under the modern Constitution would become 'Nature' and 'Society' – and the idea that only those who have the right competencies can travel between the two. Latour argues we have to get rid of both. We have to acknowledge that there exists only one Collective, an association – or rather a gathering of associations – of both humans and non-humans. And we have to acknowledge that apart from scientists, others may contribute to including nonhumans in the progressive composition of the Collective too. Hence, the tasks ahead are firstly to envision a Collective and secondly to describe an explicit proce-dure for publicly *representing* "associations of humans and nonhu-mans, in order to decide what collects them and what unifies them in one future common world" (PN: 41).

When we unpack the term 'the Collective', the first task turns out to be less grandiose than it may seem at first sight. The term is introduced to stress the *work* of collecting into a whole, that is, the work to build a common world. Immediately, Latour makes clear that he is after something quite mundane; so, we had better stop capitalizing the word 'collective'.

> The word ['collective'] should remind us of sewage systems where networks of small, medium, and large 'collectors' make it possible to evacuate waste water as well as to absorb the rain on a large city. This metaphor of the *cloaca maxima* [ancient Rome's sewage system] suits our needs perfectly, along with all the paraphernalia of adduc-tion, sizing, purifying stations, observation points, and manholes necessary to its upkeep. The more we associate materialities, institu-tions, technologies, skills, procedures, and slowdowns with the word 'collective', the better its use will be: the hard labour necessary for the progressive and public composition of the future unity will be all the more visible. (PN: 59)

Remember that Latour does not want to introduce a new institu-tion, some Parliament with a brass nameplate next to its door. He set out to *redescribe* what we have been doing all along but failed to acknowledge and to ratify – the *practices* of composing a common world. Latour is and remains an empirical philosopher. The conceptual apparatus he introduces serves the purpose of providing a language to better account for what we do and what is troubling us.

What is troubling a – or 'the' – collective? Why would a collective need a 'cosmopolitics' at all? This brings us to the second step of Latour's argument. At any moment, something can *happen* to a collective. Time and again, new entities knock on its door, offering a proposition (in Whitehead's sense of the term), asking (if not pressing) to become included. The appellant may be a virus, a new technology, immigrants who have been risking their lives to cross the Mediterranean or the US border, a species that signals that it is threatened to become extinct, a so far unidentified agent that causes a cattle-disease – anything. Whatever it is, the collective sees itself confronted with two questions. In the first place: how to take this appellant's proposition properly into account? Secondly, can we live together? The first question requires the proposition to be adequately articulated. The second question requires reflection on how the appellant would fit into the collective. Finally, the question has to be answered whether the appellant should be internalized as a member of the collective or externalized, that is, rejected.

To give appellants fair and proper treatment, to guarantee "due process", Latour formulates four requirements (PN: 102–109). The first two pertain to the first question; the third and fourth to the second one. They are addressed to the collective that sees itself confronted time and again with new propositions.

(1) You shall not simplify the number of propositions to be taken into account in the discussion. Latour calls this first requirement the requirement of *perplexity*.
(2) You shall make sure that the number of voices that participate in the articulation of propositions is not arbitrarily short-circuited. This is the requirement of *consultation*.
(3) You shall discuss the compatibility of new propositions with those that are already instituted, in such a way as to maintain them all in the same common world that will give them their legitimate place. This is the requirement of *hierarchization*.
(4) Once the propositions have been instituted, you shall no longer question their legitimate presence at the heart of collective life. This is the requirement of *institution*.

Old bicameralism

(Modern Constitution)

		House of Nature ('facts')	House of Society ('values')
New bicameralism (Cosmopolitics)	First house: taking into account	Perplexity	Consultation
	Second house: arranging into order	Institution	Hierarchization

We may note that the first and fourth requirements – roughly – cover what was formerly contained in the notion of 'fact'; and that the second and third requirements cover what was formerly contained in the notion of 'value'. Latour grants that the Moderns have correctly intuited that questions about *is* and *ought* need to be separated, but he thinks that this intuition is inadequately covered by being formulated in terms of facts (the domain of science) and values (the domain of politics). He replaces the distinction between *is* and *ought* by an alternative one (PN: 102). Where in the modern Constitution facts and values were dealt with respectively in the house of Nature (science) and the house of Society (politics, culture), he suggests that due process requires a new bicameralism, a clear division between on the one hand "the house for taking into account" (that will deal with *perplexity* and *consultation*) and "the house for arranging in rank order" (that will deal with *hierarchization* and *institution*) (PN: 153). Schematically represented, the separation of powers the modern Constitution suggested is rotated by ninety degrees (PN: 115).

Provided the four requirements are fulfilled and the separation between, on the one hand, the house of taking into account, and, on the other hand, the house of arranging into order is respected, when appellants propose themselves, they enter a procedure that will give their propositions due process. Their propositions will be taken into account (*perplexity* and *consultation*) and ranked (*hierarchization* and *institution*). If instituted in the collective, they will become a legitimate member of the collective; if not they will be externalized, rejected, but not forever. They may appeal again to be included; cosmopolitics is a continuous process, the path from perplexity, consultation and hierarchization to institution will have to be iterated time and again.

Apart from the four tasks already mentioned – (1) perplexity, (2) consultation, (3) hierarchization and (4) institution – cosmopolitics requires two other tasks, namely (5) to keep a clear *separation* between what goes on in the 'house of taking into account' and the 'house of arranging into order' and (6) to provide the collective with some self-image, a *"scenerization of the totality"* (PN 137–8), that defines (for the time being) what is inside and outside the collective.

Whatever 'scenerization of the totality' the collective will design, it will be clearly different from the image provided by the modern Constitution. The image of a Society surrounded by Nature is replaced by a distinction between what is internalized (i.e. instituted in the collective) and what has been externalized. The exteriority is no longer fixed and inert, as Nature was to Society under the modern Constitution. If an appellant is externalized from the collective, it may (and probably will) propose itself again in a new iteration of the process. So the exteriority of the collective is no longer a reserve ('Nature'), nor a court of appeal ('hard facts'), nor a dumping ground ('the environment') for Society. The outside of the collective consists of entities that may appeal to be included, and that may either threaten or enrich the collective. Latour's collectives are explicitly open to contingencies, innovation and change.

By formulating this procedure for the progressive composition of a common world of both humans and nonhumans, Latour does not propose to overthrow existing practices, to formulate an utopia or call for revolution. His ambition is to make explicit what we have been doing all the time, but implicitly, under the table.

To see this, consider, once again, the role of Pasteur in society. In the mid-nineteenth century, the variation in virulence of contagious diseases presented quite a problem. That was the problem that *perplexed* hygienists and many others and that got Pasteur started. He succeeded in identifying microbes as the cause of diseases, studied their characteristics and determined how they contributed to spreading diseases among other members of the collective – humans and cattle – (*consultation*), subsequently to suggest ideas about how to contain microbes so as to prevent them from communicating diseases (*hierarchization*). Technical facilities had to be designed, rights and burdens to be redistributed, new hygienic forms of behaviour to be facilitated and stimulated, and so on. Once that had been implemented – in cities, households, farms and in industry – the collective had both *instituted* the undeniable existence of microbes and had learned how to live better with them.

Latour's cosmopolitics turns Pasteur's practice into principle. If we focus on what Pasteur *did*, on his work, on the different *tasks* he took up, we notice that the idea that his role as a scientist was to produce knowledge that others, politicians, may use to attain their ends, only gives a very limited view on what he actually achieved. Pasteur was active in *all four* quadrants of the scheme that summarizes the cosmopolitical process. But he was certainly not the only one who did so. Politicians, hygienists, medical professionals and many others also contributed to each of these four quadrants. But each of them did so in their own way. To differentiate their roles, the fact/value distinction is too clumsy. We should discuss the different *skills* and *instruments* scientists, politicians and many others bring to the cosmopolitical process, that is, focus on the kind of *translations* they are able to perform.

This brings us to the third step of Latour's argument. Consider first the role of the sciences in progressively composing a collective (PN: 137–143). As Pasteur showed, they may contribute to each of the four quadrants. They can bring to the task of (1) *perplexity* the asset of their curiosity and open minds, plus laboratory-instruments that will allow them to detect scarcely visible phenomena and formerly unknown entities and powers. By designing suitable experiments to articulate the characteristics and powers of these entities, scientists can also contribute to the work of (2) *consultation*. Moreover, for (3) *hierarchization*, they have the competence to offer heterogeneous innovations and compromises that may help to reduce conflicts. Finally, scientists also have the knack of knowing how to (4) *institute* an entity. They know that when all is said and done, outcomes of research have to be accepted as 'hard facts'. With regard to task (5), *separation of powers*, Latour points to the right of anybody to ask their own questions in their own terms, whether they are perceived as reasonable and realistic or not by others, a right to which scientists traditionally appeal when they defend their autonomy. With respect to (6) *scenerization*, scientists can multiply the great narratives 'from amoeba to Einstein' and 'from Plato to NATO' that provide a simplified but coherent image of the common world.

Consider, secondly, the role of politicians (PN: 143–154) – the term referring here to statesmen, parliamentarians, union leaders and the like. They too contribute to each of the six tasks, but with different skills from scientists. They can contribute by being attentive to the fact that excluded entities can return to haunt the collective (*perplexity*); their role is to form concerned parties, reliable witnesses, and to mobilize stakeholders to ensure a wide variety of voices (*consultation*). They have the skill to compromise and to persuade the ones they represent to accept them (*hierarchization*) and the craft to make painful decisions, accepting the fact that building a collective also means to exclude and to make enemies (*institution*). They know that deliberation and decision-making have to be clearly distinguished (*separation of powers*). They may provide a clear sense and a narrative of who we are as a collective and what for the time being must remain outside (*scenarization*).

In *Reassembling the Social* even the 'sciences of the social' are granted a productive role in cosmopolitics. They, and in particular the 'cameral sciences', have innovated forms, standards, plug-ins and centres of calculation that may help in *consultation, hierarchization* and *institution*. Their 'panoramas' may contribute to *scenarization* by providing a "prophetic preview of the collective" (RAS: 189). The latter should however be viewed with scepticism: the concept of Society on which they rest obstructs the view on the process of cosmopolitics; and "[t]o put it bluntly: if there is a Society, *then no [cosmo-]politics is possible*" (RAS: 250).

What has been said for scientists, politicians and sociologists of the social also holds for other disciplines. Economists, for example, provide a common language allowing commensurability and calculation that may be useful in (3) *hierarchization*; while moralists stress the scruples that make it necessary to go looking for invisible entities and appellants (*perplexity*) and the need of resumption (*scenerization*).

Two other types of professionals will also be required, namely administrators and diplomats. "Administrators are going to have responsibility for distinguishing all the functions [. . .] and for coordinating the various professions" (PN: 205). Diplomats – although always belonging to one party of a conflict – are needed for their competence of sensing what the interests and moving space of the other party are. Together with administrators, they perform a seventh task, namely to ensure that due process takes its course, that the collective remains open and that the issues it is confronted with are seriously followed up.

Latour does not propose a utopia, nor does he call for a revolution. 'Cosmopolitics' is already widely practised, but political philosophy simply failed to acknowledge its existence. Pasteur's message that there are more of us than we thought has been incorporated into political practice already for more than a century. All over the world, scientists, economists, politicians and others meet in hybrid assemblies – in Brussels, in Washington and in any other capital. In contemporary politics much more is taken into account than the power, the interests and the values of people that traditional political philosophy highlights; also the power of microbes, noxious chemicals, ecosystems, and the global climate are at stake. Can all of these powers fit into an ordered, collective, peaceful existence? As Koch and Pasteur showed: it may require quite some reordering, a combination of technical, economic, social and legal measures. Today, apart from politicians and the citizens they represent, scientists, economists, jurists and moralists also bring their skills and instruments to contribute to political deliberation and decision-making, that is, to each of the four quadrants of the (cosmo-)political process. Cosmopolitics is already with us, in and outside traditional political institutions, on municipal, national, European and global levels. We only failed to acknowledge its existence properly.

In *Politics of Nature*, Latour is *redescribing* the existing practices of composing a common world with particular focus on the role of science. He redescribes the role of science *in* democracy. He is *not* pleading for 'democratization of science', that is, for including the voice of non-scientists in the production of scientific knowledge, a subject widely discussed among scholars in science and technology studies. No doubt also non-scientists can and have to contribute to cosmopolitics, for example by bringing their practical skills and experience to the job and by mobilizing new voices, locally and via the internet. But this input should not be mistaken for contributing to science; to contribute to science, specific skills, instruments and access to laboratories are required, which in most cases are out of reach for non-scientists.

This brings us to the fourth and final step of Latour's argument: the empirical evidence that cosmopolitics is already practised. In *Politics of Nature*, apart from scattered examples, the reader has to consult the literature mentioned in footnotes to get a view of the evidence. But in 2005, Latour curated *Making Things Public*, an exhibition in Karlsruhe's *Zentrum für Kunst und Medientechnologie* (ZKM), that showed an abundance of illustrations of the practices and instruments of cosmopolitics collected by Latour and his many collaborators.

The introduction to the exhibition's hefty catalogue (MTP), shows a slight change in terminology (yes, Latour keeps innovating, fine-tuning his terminology). *We Have Never Been Modern* ended with a call for 'a Parliament of Things'. Its readers may have thought that 'things' were (quasi-) objects, to wonder how speech-less nonhumans were to have any say at all, or, alternatively, whether Latour thought that apart from elected politicians, scientists should also become Members of Parliament, to represent the speechless inanimate masses, a suggestion that would immediately lead to the question how the election of these representatives should be organized. The catalogue's introduction (MTP: 14–41) removes the misunderstanding. A 'thing' is not an object, a *Ding an sich*, nor the 'things' of common sense, a *Gegenstand* (a stone, your car, etc.). The term is used by Latour in its Old English meaning, to designate a gathering, a meeting, a court, that is, any place, where a 'matter of concern' is made public. In some European countries, the old meaning has remained in use. In Norway, the parliament is called the *Storting*, in Iceland deputies gather in the *Althing*. Heidegger used the same archaic meaning to distinguish between *Gegenstand* and *Ding* (Harman 2005: 268–271). So a 'Parliament of Things' is a pleonasm, or a funny name for what in *Politics of Nature* is called the aggregate of assemblies involved in cosmopolitics. It is not one place, but a whole array of distributed ways for progressively composing a common world of humans and nonhumans by making matters of concern *res publica*.

Drawing from scientific laboratories, technical institutions, marketplaces, voting practices, churches and temples, internet forums, ecological disputes, and from practices outside the 'modern world', the exhibition at ZKM showed the variety of

> forums and agoras in which we speak, vote, decide, are decided upon, prove, and are being convinced. Each has its own architecture, its own technology of speech, its complex set of procedures, its definition of freedom and domination, its ways of bringing together those who are concerned – and even more important, those who are not concerned – and what concerns them, its expedient way to obtain closure and come to a decision. (MTP: 31)

We have never been modern; we have continued the practice of translation, creating 'hybrids' all the time. And we also have been practising cosmopolitics for more than a century. But we have been doing both "under the table", carelessly, without considering it as part of the challenge to progressively building a common world.

In *Politics of Nature*, Latour emphasizes that he has "no utopia to propose, no critical denunciation to proffer, no revolution to hope for [. . .]. Far from designing a world to come, [he has] only made up for lost time by putting words to alliances, congregations, synergies that already exist everywhere and that only the ancient prejudices kept us from seeing" (PN: 163). He has redescribed our political practice, "asking for a tiny concession: that the question of democracy be extended to nonhumans" (PN: 223), namely by explicitly acknowledging what we already do clumsily. In cosmo-politics the nonhumans are allowed to speak up, by authorizing a wide range of representatives (scientists, politicians, economists, moralists) to speak for them, that is, to articulate their propositions and to answer the question whether and how it is possible to live with them together.

Cosmopolitics is here and is here to stay. Visit any of the major institutions that help govern the world – the FAO, the World Health Organization, the OECD, the IPCC, the World Bank and the IMF, or the public authorities that protect the environment and the safety of food – or visit any ministerial department, and you will meet scientists, economists, lawyers and politicians in conference in hybrid assemblies. The forms and constitutions of these assem-blies vary; most of them are quite recently erected. And that, of course, allows us also to imagine other forms of assemblies and to take initiatives, to try to innovate and to form new ones. Once this has been acknowledged, the question can be raised how we could better institute cosmopolitics, to become more attentive to what we have been doing, to develop new competences that will be needed to progressively compose a common world. There is no grand design, nor a Providence, to guide us, nor any other authority on which we can rely. We will have to find out, empirically, experimentally.

By showing that we have never been modern *and* by asking us to ratify the procedures we have been following all the time, although only implicitly and often without the necessary diligence, Latour defends our civilization – the civilization that brought us, among others, science, the rule of law, modern ethics and politics – that is to say: provided 'civilization' is not understood in terms of modernity's 'progress', but defined "by the *civility* with which a collective allows itself to be disturbed by those whom it has nev-ertheless explicitly rejected" (PN: 208–209).

But this dual position towards modernity opens up a new ques-tion. If we need the competences of scientists, politicians, moralists, jurists and economists for this task, does that imply that we have to endorse the modern values of Science, Politics, Morality, Law, and Economy? And how should we account for these values, once the conventional accounts, based on the modern Constitution, are discarded? In *An Inquiry into Modes of Existence*, published twenty years after *We Have Never Been Modern*, Latour set out to address these questions. If not modern, what or who are we? How can we propose and defend the values we hold dear – the values "that the notion of modernization had at once revealed and compromised" – in the "planetary negotiation that is already under way over the future of the[se] values" (AIME: 17)?

6

A Comparative Anthropology of the Moderns

In 1930, on being asked what he thought of Western civilization, Mahatma Gandhi replied acerbically: "I think it would be a good idea."

How should we, twenty-first-century Western citizens, reply if we had been asked the same question posed to Gandhi in this (probably apocryphal) story? If – as Latour claims – we have never been modern, how should we present ourselves before other peoples? Which values do we consider as accomplishments of our civilization; which would we like to share with other peoples?

Whatever the answer to the last question may be, a presentation straight along the lines "we are proud to be modern, rationalized people, don't be stupid, just follow us", would not be convincing even to ourselves. Out of self-respect, we would like to express at least some of our reservations and uncertainties about Western civilization. Indeed, not all of it is a good idea. Surely, we appreciate modern medicine, science and technology, if not for enlightening us then for their substantial contributions to achieving the health and wealth we enjoy; we cherish democracy and the freedoms that are guaranteed by the rule of law; we would generously grant all of that also to other peoples. But we would also want to address some of the downsides of modernity: for example, its dominance of economic values, the exploitation – if not here, then certainly abroad – and alienation that results from worshipping Mammon. On second thoughts, we may even acknowledge that we have an ambiguous attitude to the values of modernity to which we say we are devoted. We cherish democracy, but complain about its politics; we esteem the rule of law, but hate the bureaucracy that comes with it; we respect science, but we also feel that disenchantment of the world means 'loss of meaning'. So, in private life, in search for meaning, we turn to the arts, to personal relations, and – "there really must be 'something' more than the hard facts of science, isn't there" – to spirituality in our heart of hearts.

For a minute, we may consider making a virtue out of necessity, to declaring that we are proud of our civilization because it allows us to be critical. Yes, we endorse critique. But we long for something more encompassing than the 'halved rationality' of merely calculative 'instrumental reason', namely for human interaction based on Reason, for 'communicative rationality'. Yet, in spite of the efforts of generations of critical theorists, we have trouble to explain what that would mean, if not in theory then in practice.

Adding to our confusion, in the past half-century we have come to realize that we also have to face problems of an entirely different nature: environmental degradation, scarcity of essential materials, climate change, loss of biodiversity – the effects of industrialization, mass production and consumption, of ruthlessly exploiting natural resources. Having almost exhausted the Earth and having aggressively exported modernity to other peoples, we have come to understand that to continue and to further spread the modern way of life would soon require more than one planet – two, three, five, six perhaps; the estimates vary.

So we aren't sure any longer what would be 'a good idea'. Are we confused because we still hold on to "the conviction that all the positive values in which men have believed must, in the end, be compatible, or perhaps even entail one another" (Berlin 1969: 167)? Do we only have to realize that it's all a matter of choice, that the ends of men are many and that choosing for one end inevitably involves the sacrifice of others? So is the way forward just a matter of enhancing liberalism, of freedom of choice, of better organizing our choices, which would probably come down to more modernity – more efficient markets, new technologies, and more democracy? Do we believe that by adding a little more modernity we can cure modernity's diseases – do we believe in a kind of political and moral homeopathy?

Or do we need a better diagnosis first? Are we confused because our *image* of modernity is troubled, because the *accounts* we give of the values we endorse are hopelessly inadequate, so that, as a consequence, we start messing things up when we have to present our values, consider their relations, and evaluate our civilization's achievements?

Given his earlier work, it will be little surprising that Latour thinks the latter. In *We Have Never Been Modern* he argued that the dominant account of modernity falls short. In his empirical studies he has provided alternative descriptions of science, technology and law – key institutions of modernity. By focusing on what is practically done, on the work of translation, he has found the accounts the Moderns propose to define and to justify their activities to be out of tune with their practices. For example, they claim that the methods of science allow them to directly access reality, thus forgetting all the work that is required to establish facts; to explain the binding force of law they refer to dark forces – to Nature, or to the State. In *An Inquiry into Modes of Existence*, Latour sets out to answer the questions he had left unanswered in *We Have Never Been Modern*. "If we have never been modern, then what has happened to us? What are we to inherit? Who have we been? Who are we going to become? With whom must we be connected? Where do we find ourselves situated from now on?" (AIME: 11). Latour raises these questions explicitly in the face of the ecological crisis. For Latour, at the start of the twenty-first century, we are summoned to appear not only before other peoples, but also before Gaia, "the odd, doubly composite figure made up of science and mythology used by certain specialists to designate the Earth that surrounds us and that we surround" (AIME: 9). No, in spite of what the reference to 'Gaia' may suggest, we are not entering the territory of some New Age religion. We'll get to the figure of 'Gaia' later (§ 6.5).

We need a better diagnosis first – a less confused image of modernity, a more realistic description of the values the Moderns hold dear. Latour suggests that it can be offered by a 'comparative anthropology of the Moderns' that helps contrasting different threads *within* the modern collective, rather than – what Weber did – by contrasting us, Moderns, with Them, other peoples. To allow this endeavour, a new philosophical vocabulary is required. It is introduced hand in glove with empirical observations. It has resulted in *An Inquiry into Modes of Existence*, a massive book (of almost 500 pages), presented – as one might expect from an empirical philosopher – as a "provisional report" (AIME: xix, 476).

Latour has worked on this project for about a quarter of a century, parallel to continuing his work in science and technology studies and actor-network theory. *An Inquiry into Modes of Existence* is accompanied by a bilingual website – www.modesofexistence.org

– that contains additional material, cross-references and a glossary. Readers of *An Inquiry into Modes of Existence* are invited to contribute to the website, "to extend the work [. . .] with new documents, new sources, new testimonies, and most important [. . .] [to] modify the questions by correcting or modulating the project in relation to the results obtained" (AIME: xx). Taken together, what is provided are an extensive research-protocol, a virtual philosophical laboratory, and Latour's preliminary results.

An Inquiry into Modes of Existence is not a book for the faint hearted. Readers are supposed to be well-versed in Latour's previous work and to have the stamina to devour what at first sight may appear to be bewildering conceptual acrobatics, tumbling metaphors and philosophical speculation. But once, having taken a deep breath, one dives into *An Inquiry into Modes of Existence*, one will find an enterprise of a scope and a tone rarely found in contemporary philosophy.

Although explicitly announced as a treatise in 'empirical philosophy', *An Inquiry into Modes of Existence* is written in a style that is strikingly different from most of Latour's other publications. It carries quite a heavy philosophical load and rather than presenting ethnographic details, the argument is set up by introducing a (fictitious, female) anthropologist. She proceeds by questioning the puzzles and embarrassments the Moderns experience when they try to account for values in a wide variety of domains – science, technology, religion, politics, law, fiction, their emotions, morality, and the economy. In many cases, everyday situations, rather than detailed description of practices, provide the starting point for the inquiries the fictitious anthropologist conducts. This has the advantage of circumventing both the specificities of localities and the technical details that ethnography requires. However, it does raise the question whether we are dealing here with truly empirical philosophy or with a philosopher who – by setting up a proxy – has produced suitable anecdotal evidence to defend and illustrate his ideas. Latour is aware of the danger. As an antidote, he offers his readers a modest test. They are invited to check whether the redescriptions that he – by way of his fictitious anthropologist – provides make it possible to clear up conflicts between values – conflicts that had previously given rise to more or less violent debates (AIME: 18). More specifically, in his concluding chapter he asks four questions. Can the modern experiences his fictitious anthropologist has detected on Latour's lead be shared? Does the way values have been identified allow us to respect other values?

Can other accounts be proposed? And finally: can the enquiry mutate into a diplomatic arrangement with other peoples? (AIME: 477–480)?

An Inquiry into Modes of Existence is written against the background of more than forty years of work in science and technology studies and actor-network theory. In his early work, Latour has stressed the heterogeneity of what makes up our world to conclude that to get a realistic picture we have to get rid of established distinctions that emerge from the primarily epistemological concerns of Western philosophy (e.g. the distinctions between fact and value, 'Nature' and 'Society', humans and nonhumans, word and world). In *An Inquiry into Modes of Existence* he argues for introducing new distinctions. This doesn't mean that Latour recalls his earlier work. He applies the shift in focus he had already made in *The Making of Law* to study other modern institutions. He still holds that practices are made out of heterogeneous elements and that actor-network theory is the appropriate tool to describe them. But whereas most of Latour's earlier work had focused on describing how actor-networks are *set up*, in *An Inquiry into Modes of Existence* he tries to register what is *passed* in the actor-networks that make up science, politics, law, religion, et cetera. This, he claims, will allow us to provide alternative accounts of the values that are at stake. This is what he wants his fictitious anthropologist to do.

6.1 A research protocol for a comparative anthropology

To articulate the various values that the Moderns endorse and to contrast them in a coherent and fair way, the fictitious anthropologist needs a well-filled toolbox, some new vocabulary and a research protocol. They are introduced in *An Inquiry into Modes of Existence*, closely interwoven with the question why the advent of science has made it difficult to seriously account for other values we hold dear. Themes covered in Latour's earlier work are recapitulated and sometimes amended. Rather than strictly following the book's argument, to ease understanding of the transition from Latour's earlier work through *The Making of Law* to *An Inquiry into Modes of Existence*, and to introduce the six key-terms of the latter book (*hiatus, trajectory, conditions of felicity, beings to institute, alteration* and *mode of existence*), I'll start by making up an encounter of the anthropologist with someone working in the domain of science.

We'll get to the philosophy underlying her empirical enquiry in due course.

Confusion, embarrassment, and complaints, that's what gets the fictitious anthropologist started. Suppose she encounters someone who is 'working in the domain of science'. He is a pharmacologist, developing new drugs. He proudly tells her about his work. But before long she finds him also complaining. To get funding for his research he has to spend long hours with economists from industry and with government officials; he has to confer with legal advisers because of possible patent infringements; and there is also this Ethics Review Board that requires him to do tedious paperwork before he is allowed to test his new drug on human volunteers. Is that science? Is that what a scientist is supposed to do? What a waste of time and talent.

Yes, this is what doing science involves today. No, it isn't science – this is a mixture of science, economics, politics, law and ethics. Apart from the values the Moderns associate with science – truth, objectivity, the pursuit of knowledge – there are many other values at stake. In contemporary scientific practice, many, often conflicting values have to be served.

Speaking about science as a *domain* is a major cause of the confusion. It's a wrong, cartographic metaphor; it suggests that there are clear boundaries. If we want to describe how science is *set up*, we need another system of coordinates, another topology, namely that of actor-network theory. We'll soon find that science is an arrangement of heterogeneous elements – blood samples, laboratory equipment, scientific theories, academic journals, conferences, business contracts, government grants, patent law, medical ethics. Nevertheless, what *passes through* laboratories, scientific papers and conferences, is definitely something different from what passes through e.g. a court of law. Science is a different kind of institution than law. Although the actor-network has no clear boundaries, there are some *internal* limits. If the pharmacologist wants to present his work at a scientific conference, he cannot just boast about the contracts he has received, the patent-battles he has won and how he smartly succeeded in getting approval for doing tests on human beings. At a scientific conference, he has to speak 'scientific language'. To use a musical metaphor, at the conference he has to speak in a specific *key*.

Well-said. Science comes with a specific language and ethos; it's a specific institution. So we're back at old school sociology of scientists? No. The actions of the scientists are not governed by what sociologists call the 'normative structure of science' (Merton 1973: ch. 13). As students in science and technology studies we have left that station behind us already a long time ago. Moreover, as we have just seen, in the daily practices of science also other norms and values are at stake. If we want to get an idea of what science as an institution *is*, we'll have to extract some *internal* limits of the actor-networks in which scientists participate. We have to stop talking *about* science as a supposedly well-delineated domain of activities; we need to detect the specific *key* of speaking and acting 'scientifically'. That might get us a view on what is specifically at stake in the institution we conventionally call 'science'.

Listening to a piece of music, someone well-versed in music will know the key the piece is played in and he will know when it's played in the wrong key; likewise, a music teacher may say that it *should* be played in a specific way. Similarly, a poem, for example in blank verse, may make sense only when it is read in the right way, in the right metre (Wittgenstein: 1970: 4); likewise, a 2D-drawing of a 3D-figure may be incomprehensible if we see it the wrong way, mixing up fore- and background (Wittgenstein 1969 [1952]: PU II.xi, 513). Clearly, meaning is not only a matter of how sounds, words and lines are arranged. There is some key, an explicit or implicit signpost that indicates how to interpret, play, read, or perform what follows.

How to *empirically* detect a 'key'? The fictitious anthropologist is advised to take the point where she started seriously. Just as the musician will know whether a piece is played in the right key or not, a good scientist will know whether someone is speaking scientific language or some other, for example legal, language. The anthropologist has to rely on the experience of her informants. Her pharmacologist knows very well *that* his legal advisers do not speak scientific language. However, when he tries to explain *what* the difference is – for example, when he refers to science and law as being 'different cultures' – she should become suspicious. She is advised to neglect the metalanguage in which her informants try to account for the differences they experience. To provide alternative descriptions of what science is – and what law, economics, politics and ethics are – she will have to sail on her own compass.

For a start, Austin's (1976 [1962]) philosophy of language provides useful hints. In *The Making of Law* Latour had already used speech act theory to study the specific 'regime of enunciation' of administrative law (cf. § 4.3). In *An Inquiry into Modes of Existence*, he uses Austin's notion of 'conditions of felicity and infelicity' to study what he calls *modes of existence* – a term, allowing himself some freedom (Latour 2011), he takes from Souriau (1943) and Simondon (1989 [1958]). Although this shift should not come as a complete surprise – we have seen before, for Latour 'action' has a much wider meaning than just human speech and action – the extension of the domain of application of speech act theory to (as the term 'existence' suggests) ontology needs further defence – we'll come to that later (§ 6.2 and 6.3).

What are 'modes of existence'? Amazingly, the term is never explicitly defined in *An Inquiry into Modes of Existence*. So we have to do some work ourselves. To start with, remember the 'principle of irreduction' that is the heart of Latour's philosophy and the starting-point of actor-network theory: "Nothing is, by itself, either reducible or irreducible to anything else" (IRR: 1.1.1). The world we encounter is made up by actor-networks, by actants relating, translating and defining each other. But any translation is a trial; each attempt to act or to engage in a relation can fail. In each translation a small discontinuity, a *hiatus*, needs to be practically overcome. Blood samples will get a number for future reference; their cholesterol values will be determined; the results will be included in a database; the data will be used to calculate their mean value and spread. In each translation, some transfer in space and time takes place, while form and matter change.

We need to zoom in. In the laboratory, assistants have put numbers on tubes with blood-samples; the number on the sheet will identify a blood-sample with – say – specific cholesterol values. The tubes are stored in a refrigerator; if at some point it will be needed to recheck these values, their numbers will provide access to the original tubes. Subsequently, the mean cholesterol value of dozens of blood-samples will be calculated; that value will eventually appear on one of the slides the pharmacologist will show at a conference. In each step there is a discontinuity; in each something is lost (a patient's name, age, and gender are substituted by the number that a laboratory-assistant has put on the tube containing his blood sample; dozens of data are summarized in the mean cholesterol value that the pharmacologist shows on a slide). Each step, each translation, is a risky action that may go wrong: the lab-assistant may mess up the figures; instruments may fail to function properly; calculations can be wrongly performed; et cetera. For the actor-network as a whole to function well, all the translations have to be performed well, that is, under the right *conditions of felicity*. Then, in spite of all the discontinuities, something is allowed to pass continuously along the *trajectory* of associations that are established.

What is this 'something' that is passed and maintained if the actor-network that is setup for science is functioning well? As we have seen before (§4.3), it took Latour quite some effort to detect this 'something' for law: a *moyen*. For science, his earlier work provides the relevant clue. In science, a long chain of translations allows connecting what has been done and found in the laboratory to the figures the head of the pharmacology team presents at a conference. If questioned at the conference, he will be able to go back to show that the steps that allowed him to claim that the figures on his slides *refer* to what has been found in the experiments. So let's call what is maintained 'reference'. What *is* 'reference'? It's neither language (or knowledge), nor matter (or facts). It is something specifically created in science, a constancy that is preserved when everything is functioning well, something that passes through translations, and that allows the retrieving of information, if needed.

Think about a hurdle race, Latour suggests: many hurdles have to be overcome, but something is nevertheless passing through – the runner (AIME: 107). In science, 'reference' occupies the position of the runner. If questioned, the scientist will have to show that in each of the translations the 'runner' has not fallen and that no hurdles have been knocked over. He will have to dive deep into the details of the work of his team. If the checks show nothing has gone wrong, the pharmacologist is entitled to conclude that in the long chain of translations 'reference' has passed through, that is, that all translations have been performed under the right conditions of felicity, and that the chain of translations is performing in the right 'mode' for 'reference' to be maintained.

So what is science? It is an institution that allows showing at a conference figures on a slide that refer to what has happened months ago in a laboratory or clinic, hundreds of kilometres away. It's a 'mode of existence' that allows talking about remote entities by *instituting* 'reference'.

If only the world was so simple! Then we could have continued using the old system of coordinates and could have talked about science as a 'domain'. However, in most cases, in the actor-networks we encounter more things that will have to pass through. As we have seen, the pharmacologist not only has to ensure that the figures on his slides properly refer to the phenomena registered in experiments, he has also to comply with patent-law, with the regulations of the Ethics Review Board and to satisfy the interests of industry-representatives. Different institutions, different values have to be served, different kinds of hurdles have to be overcome.

The fictitious anthropologist considers this as good news. It provides a tool for her research. She may detect *category mistakes*, namely when the key proper to one 'mode of existence' is used to address issues that belong to another 'mode of existence'. She will observe that the lawyers are not showing much interest when the head of the laboratory starts telling them at length how his team has successfully developed a new efficacious drug. He should address that speech to his scientific colleagues. The lawyers expect him to do something entirely different, namely to point out why he still believes that bringing the drug to the market will not be a violation of the patent rights of some company. To convince them, the pharmacologist has to speak legalese, their language, not the language of science.

As we noted before, when the anthropologist asks him about his unhappy encounter with his legal advisers, the pharmacologist will probably talk about the different 'cultures' of law and science. He might explain that scientists and lawyers have different *views* on one and the same thing, that is, the drug that has been developed. While the scientists *see* the drug *as* an efficacious medicine, lawyers *see* the same drug *as* a possible violation of patents.

By conceiving the drug as a substance on which different *views* are possible, the pharmacologist falls back on the epistemological language and the worldview of the modern Constitution that suggests that on the one side there is Nature, matter, substance, which is universally given, while on the other side there is Society with its plurality of cultures and perspectives. However, as readers of *We Have Never Been Modern* know, the modern Constitution and the language it suggests is hopelessly inadequate to account for modern practices, including science. This is not the route Latour wants the fictitious anthropologist to take. She is advised to ignore the pharmacologist's metalanguage to turn to the language of being, to ontology.

Should she speak about 'hybrids', or perhaps about 'quasi-objects' or 'propositions' (cf. § 5.3)? Well she could do so. But Latour has finally found better terms to express what he is up to. On his lead, the fictitious anthropologist is advised to distinguish between *being-as-being* and *being-as-other*. Being-as-being "seeks its support in a *substance* that will ensure its continuity." In contrast, beings-as-other "depend not on a substance on which they can rely but on a *subsistence* that they have to seek out at their own risk" (AIME: 162). To continue to exist, beings-as-others have *hiatuses* ('hurdles' in the metaphor suggested above) to overcome. Time and again, they have to face *trials*. They will continue to exist only when they stand up to these trials.

Terrible jargon, indeed. (Do the French equivalents, *être en tant qu'être* and *être en tant qu'autre* sound better?) But in the light of the history of ideas, introducing the distinction is a bold move indeed. To appreciate this move, a comparison with the Darwinian revolution in biology may help.

Darwin introduced a new logic of reasoning in biology. He shifted biological thinking from essentialism to population thinking (Mayr 1982: 45–47). Before Darwin, a biological species was conceived as a group of individuals that share some essence. What decides that an animal is a horse (rather than – say – a donkey) was conceived to be the 'Idea of a horse' (Plato's *eidos*), a definite form that it and other horses share. Of course, it was known that not all horses look alike; but the individual variations within a species were taken for imperfections. Behind the variety of appearances, true reality was supposed to be made up of stable, eternally given essences. Emphasizing the fixed and the final, variety and chance were perceived as disturbances, as defects and unreality.

Darwin radically broke with this mode of thinking. After Darwin, biologists conceive a species as a population of individuals that may interbreed; in other words, a species is a *population* of individuals separated from other populations by a reproductive gap. Continuation of a species' existence depends on how successful the individuals in this population are in reproducing. Without sufficient offspring, a species will become extinct. The consequences of this conceptual shift are far-reaching. Variety within and among species is no longer conceived as imperfections from eternally given essences and change no longer conceived as divergence from some stable, given design of the animal kingdom. They are accounted for in terms of processes of variation and selection, of reproductive success.

Latour extends the logic of reasoning Darwin introduced in biology to *any* being, including inanimate ones. Of course, he is certainly not the only contemporary philosopher to take issue with the long tradition of essentialism. Anti-essentialism has become a major issue in twentieth-century philosophy (e.g. Rorty 1989). Its sources are diverse: philosophy of language (e.g. the later

Wittgenstein), Heidegger's philosophy, and discussions about rationality after Kuhn. But in most contemporary philosophical debates, Darwin plays at best a minor role. In contrast, in early twentieth-century philosophy it was explicitly recognized that Darwin's work would have to have major consequences for philosophy (e.g. Dewey 1997 [1909]). His influence is noticeable in Bergson, in pragmatism, in Whitehead and in Popper, who applied Darwin's logic of reasoning to epistemology and the philosophy of science (Popper 1972). Also Tarde – who had read Darwin and who conceived everything that exists to be a 'society' (Tarde (1999 [1895]: 58) – can be considered as a radical population thinker. Among them, Whitehead stands out.

In his 1925 Lowell lectures, published as *Science and the Modern World*, Whitehead outlined "the essentials of an objectivist philosophy adapted to the requirements of science and to the concrete experience of mankind" (Whitehead 1967 [1925]: 89). Whitehead argued that the cosmology that had reigned in science for three centuries and that presupposes the ultimate fact of an irreducible brute, senseless, valueless, purposeless matter, was no longer suited to the scientific situation that [in 1925] had arrived. As an alternative theory of nature, he proposed an "organic theory" that takes organisms, rather than matter (substance) as its starting point.

According to Whitehead, the metaphysics that had been leading science for centuries suffers from what he famously called the 'Fallacy of Misplaced Concreteness', that is, the expression of concrete facts under the guise of abstract logical constructions (Whitehead 1967 [1925]: 50–51). Whitehead argued that when physicists speak about matter, they speak about something that is already defined in space and time; what they conceived of as matter, is reality *plus* a specific way of formatting and localizing events and entities. What is left out of the picture are the instruments that allow this formatting. According to Whitehead (1967 [1925]: 79) "the whole concept of materialism only applies to very abstract entities, the products of logical discernment." As an alternative to materialism, Whitehead suggested that the concrete enduring entities are organisms, that is, entities that have to survive – to change and endure – to exist.

What allowed us to see the limits of the materialist worldview is the emergence of new instruments, in science (e.g. Micholson's experiments, Whitehead 1967 [1925]: 114 ff.) and in the arts. "A fresh instrument serves the same purpose as foreign travel: it shows things in unusual combinations. The gain is more than a mere addition; it is a transformation" (Whitehead 1967 [1925]: 114). Whitehead argued at length that taking the shift from a cosmology that is based on the concept of matter to one that is based on the concept of organism would not only resolve the puzzles the theory of relativity and quantum theory present us with, but also brings our conceptual apparatus in line with the experience of nature articulated among others by the arts. To avoid misunderstanding: Whitehead was not pleading for vitalism. Vitalists continued to accept materialism for inorganic nature, to introduce a distinction between living creatures and inanimate ones. Whitehead erases this distinction, to apply the theory of organism to nature *tout court*.

In Whitehead, Latour found a philosopher who had articulated many of his own intuitions. But in contrast to Whitehead, Latour is neither after a theory of *nature* (a concept that suggests a distinction with society), nor concerned with the conceptual problems of quantum mechanics and Einstein's theory of relativity. He is in search for a philosophy that is flexible enough to allow his fictitious anthropologist to do her work, one that allows conceiving a *plurality of beings* that for the continuance of existence have to *subsist*, rather than one that only conceives a plurality of *views* on one and the same *substance*. Introducing the concept of being-as-other (replacing the term 'organism' Whitehead introduced in his theory of nature), he observed that

> [t]hanks to Darwinism, we have been familiar for a century and a half with the risk taken by the entities that thrust themselves into *subsistence* through the intermediary of reproduction. [. . .] We have finally understood that there was no Idea of a Horse to guide the proliferation of horses. Here ends, on this point at least, the quarrel of the Universals. (AIME: 102 italics added)

To continue to exist, to subsist, beings-as-other do not rely on a substance (essence); they have to overcome hiatuses. How? Interbreeding may be the preferred way for living creatures, but more generally formulated the answer is: by entering into (any kind of) relation with other beings-as-other, namely by translating other beings-as-others, or by being translated, to form an association that is stronger, that is, has better chances to withstand trials. Does that mean that there is no difference between living organisms and inanimate nature? Of course not. But this difference has to show in the translations that are performed, in the conditions of felicity, in what is passed on and is instituted.

Introducing the concept of 'being-as-other', however, leads to an even more radical anti-essentialist idea. Depending on the kind of translations a being-as-other performs or is entered into (that is, depending on the actor-networks in which it becomes included), it may continue to exist *in different ways,* in different 'modes of existence'. Shifting to the concept of being-as-other allows us to speak about the different hurdles (hiatuses) a being has to overcome to continue existence and about the different events that may happen to a being to affect its existence. Introducing the notion of being-as-other allows investigating "how *many other forms of alterities* a being is capable of traversing in order to continue to exist" (AIME: 163).

To get the idea, rethink the pharmacologist's situation. If 'reference' is properly instituted and preserved, the pharmacologist can speak at a conference about the efficacious drug his laboratory has developed many months ago, hundreds of kilometres from where he is standing now. That is the specific *alteration* science allows: to speak and write about remote entities – in this case: about the existence of a new efficacious drug. But for his drug to become a marketable medicine, hurdles of a different kind have to be overcome too. If the production of the drug violates the patent some company holds, the drug cannot be brought to the market (without first negotiating an agreement with the patent-holder, something that those who invested in the research will deplore). In short, to exist as an efficacious *and* as a marketable drug, the actor-network in which the drug is produced has to allow different forms of alteration; not only 'reference' has to pass through. Also something different has to pass in the actor-network, something specific for the institution of (patent) law.

While the pharmacologist talks about one and the same substance (being-as-being) that can be viewed and interpreted in different ways, Latour's language allows his anthropologist to speak about a marketable and efficacious drug as a being that has a different kind of *existence* than a drug that is either not efficacious or not marketable. A drug that succeeds in overcoming the various hurdles patent law presents will have a different existence, a different 'career', than one that fails in taking these hurdles. That difference is not a matter of diverging 'views' and 'interpretations' on one and the same substance, but shows that they are different existents. The one *is* a marketable commodity; the other one isn't.

By teaching his poor fictitious anthropologist the distinction between being-as-being and being-as-other, Latour has made two moves at once. In the first place he allows her to distance herself from essentialism and the epistemological language that is implied in the pharmacologist's explanation of what is at stake. In the same move, Latour has given (individual) *existence* pride of place over *substance* (essence) and he has moved *pluralism* from the level of language, culture and society to the level of ontology. He has introduced population-thinking on the level of ontology.

The fictitious anthropologist's toolbox is filled up. Now the various terms introduced in the preceding discussion allow us – at last! – to define the term 'mode of existence'. A *mode of existence* is defined by specifying five characteristics:

(1) The *hiatuses* that need to be overcome;
(2) The continuity it allows through associations, the *trajectory*;
(3) The *conditions of felicity* of these translations;
(4) The beings that are *instituted* and passed through;
(5) The *alteration* of beings-as-other it allows.

While filling her toolbox, the anthropologist has got an idea how to specify each of these five points for *science* as a mode of existence. Summarized in a few phrases, they are (AIME: 488–489):

(1) *Hiatuses*: Distance and dissemblances of forms (e.g. for later reference, blood samples are assigned numbers).
(2) *Trajectory*: Paving with inscriptions (they are the means to overcome the hiatus of form, allowing information about blood samples (e.g. their cholesterol values) to be translated downstream the chain of translations up to the point where the pharmacologist can speak about the efficacy of the drug on a conference).
(3) *Conditions of felicity/infelicity*: Bring back/lose information (the translations should allow access to previous steps, to bring back information that is lost in the translations).
(4) *Beings to institute*: 'Reference' (constants through transformations).
(5) *Alteration*: Reach remote entities (allowing the pharmacologist to speak about a drug developed in his laboratory, at a conference, or in a scientific publication).

Our fictitious anthropologist now knows what she has to do. Instead of joining her informants in talking about different 'interpretations' and conflicting 'cultures', she has to start 'an enquiry into different modes of existence'. She has found a way to study the institutions the Moderns claim to value in terms of different 'modes of existence', each having a specific *key*. She knows that to detect this key she has to focus on what is maintained in the chain of translations in which beings (-as-other) will be *alterated*, on what will be *instituted* and is passed in a *trajectory* through an actor-network if the correct *conditions of felicity* apply. This, she hopes, will provide her with a way to better articulate what each value stands for: a specific 'regime of enunciation', a specific 'mode of existence'. Yes, we set up actor-networks by arranging heterogeneous elements; that may confuse us. But with the proper vocabulary and research protocol at hand, the anthropologist may try to *empirically* detect what is maintained, passed through, to register the values of the institutions that are involved.

Having learned the research protocol, she can now start trying to redescribe also other activities that the Moderns identify in terms of 'institutions', 'values', 'domains', 'perspectives' and 'cultures': law, religion, fiction, et cetera. After the meeting with the pharmacologist, she will meet people who talk about practices that do not seem to refer to any 'substance' at all – people who read novels, listen to music, enjoy art, who are politically active and maybe go to a church on Sundays. Yes, they say they are moved by what they read, hear and see, but they have no way to account for *what* is moving them. Confused, they speak vaguely about 'feelings'. But nevertheless, they *experience* that something is moving them, and has some force on them – they note for example that *this* piece of music *should* be played in a specific way. The piece seems to *ask* for it – how come?

The introduction of the concept of 'modes of existence' allows Latour to account for the values the Moderns hold dear in *ontological* terms to open up an enquiry into the *variety* of modes of existence. Under the modern Constitution, the exclusive position of science – as the unique way to access Nature – condemns other experiences (especially of fiction and religion, as well as emotions) to be second-rate fantasies without some serious reality-check. But in Latour's philosophy, the unique position science is allowed under the modern Constitution has disappeared. Science, law, fiction, religion, are all treated on a par; they all are redescribed by the same protocol, to become articulated as different modes of existence. Introducing the vocabulary of *An Inquiry into Modes of Existence* enables *contrasting* these modes of existence, to register the worth and the form of eloquence of each *in their own terms*, rather than *comparing* them either by taking science as a standard that none of the others can meet, or by relativizing all of them by calling them 'cultures' and thus losing what science – and each of the other institutions – makes unique. So no, Latour does not *relativize* science, he does not condemn science to be 'just another form of belief', 'another culture'. He is not a postmodernist, not a relativist. He allows a variety of modes of existence to be investigated *each* on *their own terms*. His ontology does not downgrade the value of science, but upgrades the other values that the Moderns say to hold dear. They are allowed to be articulated each in their own terms.

So there she goes. The fictitious anthropologist notes that in accounting for their values and institutions the Moderns are confused. Annoyed, they talk about different 'cultures' allowing different, often conflicting views on one and the same 'substance'. Utterly confused, some of them even consider science as just another 'culture', a poor one indeed, one that for its 'halved rationality' has no way to account for all the other values they hold dear. When it comes to account for their values the Moderns are at a loss.

The fictitious anthropologist sets out to get a better image of modernity, an alternative account of its values and institutions. She wants to know how to *speak well* of *each* of them. She will start with registering the embarrassments, confusions and complaints she hears, and will detect category mistakes. She has learned her lessons, her toolbox is full: she has acquainted herself with the concepts of a key, hiatus, trajectory, conditions of felicity and infelicity, of beings to institute and alteration, and she knows that time and again she will encounter *crossings* of modes of existence. She is ready to enquire into different modes of existence. How many are there, what are they, how do they relate to each other? That's what she wants to know.

Once she has identified them, she thinks that the Moderns will be less confused and in a better position to appear before the others, and before 'Gaia' – to discuss in what sense and to what extent 'Western civilization' is a good idea. Perhaps a discussion about which values have to be re-instituted has to be started. Do we have to recall modernity, she wonders, to repair errors that originated in the long history of modernity and that – if we continue along its path – will cause substantial risks to ourselves, to the others and to the one Earth we all share?

The anthropologist hopes that her work may provide a platform for discussing these questions with some self-respect. No, she does not *criticize* modernity. Her aim is to provide a better view on what the Moderns conceive as modernity's contributions to civilization, to end their *aporia*. Coming from the modern world she may act as a *diplomat* offering other peoples her letters of credence: this is our civilization, what about yours?

6.2 'Empirical philosophy' redefined

An Inquiry into Modes of Existence aims to provide a better, richer view on our values, on who we are and how we live, and to provide a platform for a fair diplomatic exchange with other peoples. Latour wants us to get that view not by speculation, but to develop it by way of an empirical anthropological investigation.

The turn from epistemology to ontology is a common thread in Latour's work. In previous works, he defended this turn by pointing to ethnography; we might say that he argued for ontology on empirical grounds. In *An Inquiry into Modes of Existence* Latour sets up a *fictitious* anthropologist and he appeals to the reader's common sense, inviting them to check whether the redescriptions that he – by way of his fictitious anthropologist – provides make it possible to clear up conflicts between values or not. Is this still 'empirical philosophy'?

Are we nitpicking? No. The question becomes more pressing when we realize that Latour appeals to a much wider notion of experience than is usually covered by this term. For example, when he claims that in a mode of existence something is instituted that can be *experienced*, he is certainly not referring to something an empiricist would accept as 'empirical evidence'. 'Reference', for example – to use the example discussed above – would certainly not qualify for this; nor would *'moyen de droit'*, singled out in *The Making of Law* as what is passed through in law. Moreover, when Latour comes to discuss people experiencing 'religious beings', most empiricists will be puzzled – if not leave the room. They might be willing to say that people imagine these beings, fantasize about them, or have intimate personal feelings – but a 'religious being' is not something that qualifies to be *empirically* registered in the way we may see a mountain, touch the table, or hear a sound, is it?

The problem, according to Latour, lies entirely on the empiricist's side. That doesn't mean that he wants to make room for supra-natural experiences, or has become a mystic. Referring to William James (1996 [1912]), he claims 'experience' to extend beyond what classical empiricism (Latour calls it 'first empiricism') understands by this term.

In *Essays in Radical Empiricism*, James (1996 [1912]) argued that Locke, Berkeley, Hume and Mill, the classical empiricists, defended an amputated concept of experience that only led "to the efforts of rationalism to correct its incoherencies by the addition of trans-experiential agents of unification, substances, intellectual categories and powers, or Selves" (James 1996 [1912]: 43). The result was a long history of what James considered to be fruitless philosophical discussions. Conceiving experience in terms of discrete, atomic units, the classical empiricists have missed what James understood as the most important fact about it, namely that experience is a continuous stream, each part having no distinct boundaries, each leading to and compenetrating the next.

As a consequence of conceiving experience as a continuous stream, James (1996 [1912]: 42) argues that "the relations that connect experiences must themselves be experienced relations, and [that] any kind of relation experienced must be accounted as 'real' as anything else in the system." A simple example may illustrate the point. Suppose we are in Shakespeare's birthplace. A classical empiricist will say that we see Stratford, the city, and that we see the river Avon, but that their relation, the preposition 'upon' in 'Stratford-*upon*-Avon', is supplied by our reasoning, by our mind. According to James we also *experience* this relation. Walking through the city, encountering the river, we *experience* (rather than conclude, i.e. reason) that Stratford is *upon* Avon.

Allowing relations (in particular prepositions) to be experienced, James criticizes – not to say: ridicules – both empiricist and rationalist philosophy. Like Latour, he abandons the subject/object partition and its derivatives. The supposed bifurcation between on the one hand the world that can be experienced and on the other hand our mind, consciousness or language, leads us astray. Dualisms – such as between the knower and the known, subjective and objective, mental and physical, thought and thing, value and fact – do not designate unalterable or fundamentally disparate metaphysical categories, that is, different kinds of stuff. According to James, they are tools for designating *functional* distinctions among elements of the stream of experience. First, there is experience; "it is made of *that*, of just what appears, of space, of intensity, of flatness, brownness, heaviness, or what not" (James 1996 [1912]: 27). Then we may start to sort our experiences functionally, for example by introducing a distinction between what is attributed to the knowing subject and what is attributed to the known object.

Pure experience, according to James (1996 [1912]: 93) is "the immediate flux of life which furnishes the material to our later reflection with its conceptual categories [. . .] a *that* which is not yet any definite *what*, tho' ready to be all sorts of whats [. . .]." The 'whats' may be minds and bodies, people and material objects, values and facts, but what they are depends not on a fundamental ontological difference among these 'pure experiences', but on the *relations* into which they enter and the function attributed to them. One and the same experience may have different functions. For example, if we observe a shoal of fish in a river, we may conclude that we have seen fish (the fish being the object of what we see); but we may also observe that the quality of the river-water has improved (using the appearance of fish as an indicator for water quality, that is, conceiving it as a tool, an extension, of the knowing subject).

As a consequence, for James, values are as perceptible in experience as facts are. In contrast to the platitude that beauty is in the eye of the beholder, James emphasizes:

> We discover beauty just as we discover the physical properties of things. Training is needed to make us expert in either line. Single sensations also may be ambiguous. Shall we say an 'agreeable degree of heat', or an 'agreeable feeling' occasioned by the degree of heat? Either will do; and language would lose most of its aesthetic and rhetorical value were we forbidden to project words primarily connoting our affections upon the objects by which the affections are aroused. (James 1996 [1912]: 143–144).

We say "[t]he man is really hateful; the action really mean; the situation really tragic – all in themselves and quite apart from our opinion"; we "talk of a weary road, a giddy height, a jocund morning or a sullen sky" (James 1996 [1912]: 144). It's the road that *is* weary, the sky that *is* sullen – that's how we experience them.

By endorsing James' concept of experience in *An Inquiry into Modes of Existence*, 'empirical philosophy' gets a different meaning. From now on, an 'empirical philosopher' is not just someone who likes to explore and back-up his philosophical claims by empirical, ethnographical or historical investigations. 'Empirical philosophy' has come to denote a philosophical attitude and a metaphysical position that identifies what is real with what is experienced. With

James, Latour declares: "we want *nothing but* experience, to be sure, but [also] *nothing less* than experience" (AIME: 178). Empirical philosophy begins and ends in experience; but it acknowledges that in between a lot of thinking, sorting and work may have to be done.

> What might be called the *second empiricism* (James calls it *radical*) can become faithful to experience again, because it sets out to follow the veins, the conduits, the expectations, of relations and of *prepositions* – these major *providers of direction* [i.e. *keys*]. And these relations are indeed in the world, provided that this world is finally sketched out for them – and for all of them. Which presupposes that there are beings that bear these relations, but beings about which we no longer have to ask whether they exist or not in the manner of the philosophy of being-as-being. But this still does not mean that [as Husserl suggested] we have to 'bracket' the reality of these beings, which would in any case 'only' be representations produced 'by the mental apparatus of human subjects.' The being-as-other has enough declensions so that we need not limit ourselves to the single alternative that so obsessed the Prince of Denmark. 'To be or not to be' is no longer the question! (AIME: 178)

Allowing relations (in particular prepositions) to be *experienced* takes us away from traditional philosophies of being. Experience provides a *key*: it prepares us for what follows. Having experienced Stratford *upon* Avon, we'll perceive the city and will talk about it in a different key. No, it's not only our perception that has changed. Next time we go to visit Shakespeare's birthplace, we go to 'Stratford-upon-Avon'. The city has changed; it has become another being. But to allow this conclusion, we must suppose it to be a being-as-other, rather than a being-as-being. The existence of 'Stratford-upon-Avon' does not rely on a substance, but depends on subsistence. If for some reason the Avon would disappear, Stratford would become a different city. Walking through the city, we would still be in Shakespeare's birthplace, but no longer in 'Stratford-upon-Avon'.

Latour's move to 'second empiricism' has consequences for some of the themes discussed in his earlier works. As we have seen before, 'construction of facts' has been a major topic in Latour's early work. The idea has often been met by disbelief, up to the point at which people started thumping a table to show that reality 'really exists'. To that critique, we have our response ready: that something is constructed does not exclude that it's real: the table on which you are knocking to make your point is both real and constructed, isn't it? Another question is harder to answer: why is it so *difficult* to construct a fact? Why does it requires so much effort, why does *not* 'anything go', why do we fail so often?

To answer these questions, Latour notes that the use of the term 'construction' is confusing us. The term tries to say three different things at once (AIME: 158–9; cf. also RAS: 88–93).

In the first place, 'to construct' means to make something happen, to make (someone or something) do (something) – the point nicely covered by the French expression *faire faire* ('make do'). The engineers who have constructed a machine *make* it *do* what it's meant to do. "Every use of the word 'construction' thus opens up an *enigma* as to the author of the construction: when someone acts, *others* get moving, *pass into action*" (AIME: 158).

In the second place, "to say of something that it is constructed is to make the *direction* of the vector uncertain." Latour cites Balzac, who is the author of his novels, but who writes that he has been "carried away by his characters", who have forced him to put them down on paper. So who is acting? Who is making who to do something (AIME: 158)?

In the third place, Latour notes that "[t]o say of a thing that it is constructed is to introduce a *value judgement*, not only on the origin of action [cf. the two points before] but [also] on the *quality* of the construction" (AIME: 159). To say that something is constructed is to say that it is constructed *well*. And this, of course, is what may keep a novelist awake at night: is this character adequately constructed? Similarly, a scientist may worry: is what I have constructed in my laboratory a fact or an artefact?

Because the term 'construction' tries to say too much at once, Latour concludes that we had better abandon the term. Souriau (1943) introduced a term – *instauration* – that arguably covers better what 'construction' meant to say.

> An artist, Souriau says, is never the creator, but always the instaura-
> tor of a work that comes to him but that, without him, would never
> proceed toward existence. If there is something that a sculptor never
> asks himself, it is this critical question: 'Am I the author of the statue,
> or is the statue its own author?' We recognize here the doubling of
> action [first point above] on the one hand, the oscillation of the
> vector [the second point] on the other. But what interests Souriau
> above all is the third aspect, the one that has to do with the quality,
> the excellence of the work produced: if the sculptor wakes up in the
> middle of the night, it is because he still has to let himself do what
> needs doing, so as to finish the work or fail. (AIME: 160)

Souriau's notion of instauration has the advantage of articulating all three features: the double movement of *faire faire*; the uncertainty of who or what is the author; and the risky search, without a pre-existing model, for the excellence that will result (provisionally) from the action. It covers the *experience* of novelists and scientists better than the term 'construction'.

The shift from 'construction' to 'instauration' is more than an innocent semantic subtlety. Having introduced 'instauration' Latour immediately draws an important, but also quite enigmatic conclusion:

> [T]he act of instauration has to provide *the opportunity to encounter beings being capable of worrying you.* Beings whose ontological status is still open but that are nevertheless capable *of making* you *do* something, of unsettling you, insisting, obliging you to speak well [. . .]. *Articulable* beings to which the instauration can add something essential to their autonomous existence. Beings that have their own resources. (AIME: 161)

How can beings – "that have their own resources" but that still need to be articulated further (for example a character in a novel that is still unfinished, or a fact that has not yet been established) – make you (a novelist, a scientist) do something, unsettle you, insist and even oblige you to speak well about them? How can these beings have a grip on you? What is the source of this normative force? We need to unpack the point Latour makes.

Why are the scientist and the novelist waking up at night? What are they worrying about? Are they afraid of becoming the laughing stock of colleagues when they have made a mistake? Or is it because "the responsibility of the masterpiece to come – the expression is [. . .] Souriau's – hangs [. . .] on their shoulder" (AIME: 160)? Does the fact to be stated or the character that has to make an appearance in the novel appeal to them to act responsibly? The first option would not trouble any philosopher (they know the fear of becoming a laughing stock all too well); but as the above quote makes clear, Latour goes for the other option. What is troubling the scientist, and the novelist at night is not their position in some coterie. So what is worrying them? How can something they still have to 'instaurate' have a normative force over them?

To answer this question, we may go back to Austin's speech act theory. Provided the appropriate conditions of felicity are fulfilled, by uttering 'I promise to pay you €100 tomorrow' Peter has made a promise; he has made a commitment. He ought to pay Mary €100 tomorrow. Both Peter and Mary know this; being competent speakers of English, having internalized the constitutive rules of the English language, they both know that uttering this sentence *counts as* making a commitment to pay the money the next day.

For philosophers of language, interested in explicating meaning and the force of speech, this will do. But for a philosopher "who wants *nothing but* experience, to be sure, but [also] *nothing less* than experience" and who defines reality in terms of experience, it doesn't. How does Peter *experience* that his words commit him to do tomorrow what he promised? What makes this commitment *real*?

The words Peter has uttered constitute a *relation* between his act – his uttering these words – and what has to be done tomorrow. In James' philosophy, we are entitled to say that Peter *experiences* this relation. If questioned, he will sort it out in common terms: yes he knows that tomorrow Mary will expect him to deliver the money, and he knows that Mary will be disappointed if he fails to do so; she may even become quite angry or upset. So in Peter's experience, Mary is already included, but only in vague contours: when uttering his words, Peter does not yet know how Mary will react if he fails to keep his promise. So in Peter's *experience*, Mary is a "being whose ontological status is still open" – she may turn out to be either an understanding person or an angry one. But Peter experiences her nevertheless as capable of making him do something, of insisting, obliging him to pay the money he has promised her.

Now note that not only speech acts but also other actions may invoke a similar experience. Writing a novel? As Balzac experienced, that implies a commitment to that novel and to its characters. Being engaged in scientific research? Again, that comes with an experience: you have to care for 'reference'; you know that you have to be careful when making inscriptions and performing calculations: between your translations some constancy has to be preserved. So you worry not (or at least: not only) because you fear to become a laughing stock if you don't live up to what is expected from you. Having uttered a promise, you know that there is someone waiting for the promise to be kept; writing a novel, you know that there are characters waiting to be properly put on paper; as a scientist, you know that inscriptions have to be handled and calculations to be made in a way that preserves reference. While you are acting, that is part of your *experience*. You don't yet know how to describe the character well; you don't yet know what the fact you want to establish will be. So you worry: how to speak and act *well*, how to articulate these beings whose ontological status is still open but who appear to you to have resources of their own? You know that if you fail, you will instaurate an artefact, or a character that is flat and unrealistic. That is why you worry. You experience encountering "articulable beings to which [your] instauration can add something essential to their existence." They have some normative force.

In modern philosophy, facticity and normativity are contrasted. The contrast is elegantly summarized in a phrase of the nineteenth-century philosopher Lotze: "*Seiendes ist, Werte gelten*" (cited in Schnädelbach 1983: 199) – 'the existent *is*, values have *validity*'; for once the German language is more succinct than English.

Normativity is a child of Kant's philosophy. Having separated the realm of thinking (*res cogitans*) and the realm of things (*res extensa*), Descartes had to address the epistemological problem how we can be sure that mental concepts truly represent the world outside. How to be certain that what is represented is real, rather than an illusion? Descartes' answer was: if it stands before us *clare et distincte*. For Kant, Newton had shown that we can attain true knowledge about the world. He asked transcendental questions: how is that possible? What does it imply for what we call 'knowledge', 'thinking' and for what we call 'the world'? To answer these questions Kant introduced a distinction between directly observing something and having knowledge. Brandom summarizes Kant's position with admirable efficiency:

> Kant's big idea is that what distinguishes [human] judgement and action from the responses of merely natural [nonhuman] creatures is neither their relation to some special stuff [*res cogitans*] nor their peculiar transparency [*clare et distincte*], but rather that they are what we are in a distinctive way *responsible* for. They express *commitments* of ours: commitments that we are answerable for in the sense that our *entitlement* to them is always potentially at issue; commitments that are *rational* in the sense that vindicating the corresponding entitlements is a matter of offering *reasons* for them. (Brandom 2000: 80)

While Descartes addressed our grip on the concepts of cognition, Kant addresses *their grip on us*: when claiming knowledge, when using a concept, to what does it *commit* us? For Kant, knowledge was a normative concept. Having established that, it opened the path to investigating other forms of normativity, to other concepts and actions where accountability is at stake, questions can be raised and reasons may have to be provided.

Having left epistemology for ontology, for a long time Latour ignored normativity as an issue. He provided alternative *descriptions* of what doing science and technology involved. Writing most of the time about the work of successful scientists – like Pasteur and Guillemin's people at the Salk Institute – what they were committed to was taken for granted. That changed, when Latour turned to analysing a failed project, *Aramis*. It's failure is exposed in terms of Aramis, a fragile actor-network, not being 'loved' enough. However, except for words that Aramis, if he would be able to speak, would have said (AR: 293–296), that lack of commitment is left unanalysed. In *An Inquiry into Modes of Existence*, Latour raised the bar. Expanding – on James' lead – the notion of 'experience', he incorporates normativity into his ontological project. He speaks about responsibility (for a character in a novel, for a fact – for "beings whose ontological status is still open"). Not only our concepts and actions have a grip on us; other *beings* may have a grip on us as well. They may worry us, keep us awake at night, oblige us to speak well about them and to treat them well. What beings do the Moderns encounter in each mode of existence and how to veritably speak well about them, that's the question the anthropologist has to answer.

6.3   Enquiring modes of existence

While filling her toolbox, the fictitious anthropologist has gradually come to understand how to redescribe the specific value *internal* to science. Its mode of existence is baptized [REF], for 'reference'. Latour will use similar codes for other modes of existence. In *The Making of Law* he had already studied law. Its analysis guides us to its mode of existence [LAW]. The anthropologist is now ready to investigate other modes of existence. One of them is [REL], for religion.

Does religion stand for a value that the Moderns endorse? For one thing, the anthropologist will find abundant traces of religion in their joint European heritage, their languages and arts. But religion also divides them; it is the subject of heated debates. There are Moderns who consider themselves to be religious people, who think spirituality rings true – although instead of regularly visiting a church they may, reluctantly, opt for a long weekend retreat in some monastery. Other Moderns consider religion to be a relic from the times before the Enlightenment, before the advent of science. They wonder why anybody would take the old scriptures and rituals seriously. Perhaps the Moderns' attitude toward religion is best captured by the sardonic inscription that Alexander Kinglake, a nineteenth-century English travel writer and historian, suggested should be placed on all churches: 'Important if true' (cited in Skinner 1992: 133).

But whatever their personal view of religion, when opening their morning paper, the Moderns will discover another reason for developing a clearer view on the subject. They are confronted with the fact that religion matters to many people – up to the point that they are prepared to fight and to die for it. What moves them? Should one look for psychological, sociological, historical, or economic explanations?

Explanation is not what our anthropologist is after. "There is nothing 'behind' religion – no more than there is anything interesting 'behind' fiction, law, science and so on. [. . .] The social consists of all of them together" (AIME 307). She wants to describe what religious experience *is* and what it means to speak religiously. She is not so much interested in the truth of religious speech, as in its truth conditions.

What is the problem with specifying religious experience? Conventionally, we sort ages by talking about 'before' and 'after Christ'. When talking about religion, 'before' and 'after the Scientific Revolution' would be more appropriate (AIME: 319 ff.). The successes of science invited religious people to answer the question: 'what do *you* have to say about remote things?' The advent of science turned religion into a 'belief in something', in God, in something inaccessible, at least until death has come.

When the epistemological accounts of science removed from sight all the translations, all the work that science requires to access remote entities, to suggest that scientists, by having rational minds and methods, have direct, unmediated access to them, even the bar for religion was raised. Using the metaphor of the computer-mouse, by calling it 'Double Click communication', Latour ridicules the idea that straight unmediated talk might give us access to reality without overcoming any hiatus at all. The idea of 'Double Click communication' provides an amputated description of science; for religion, Latour claims it completely loses what it was supposed to grasp.

Latour's critique of what he came to call 'Double Click communication' dates back from his student days, when he inverted

Bultmann's attempt to demystify the New Testament to open up the Gospel for rationalized people (cf. ch. 1). While Bultmann tried to point to the truth of Christianity by eliminating everything that couldn't genuinely be attributed to the historical Jesus, Latour suggested that the truth conditions of the Gospel resided in the long chain of continuous inventions by later interlocutors. In *Rejoicing – Or the Torments of Religious Speech*, his most personal book, he recapitulates his point (REJ: 94 ff; 129 ff).

> [The] conveyance towards the 'historical Jesus', which has its own usefulness, its particular grandeur, its specific demands, its rigour, its seriousness, can't be confused with the usefulness, the grandeur, the strictness, the rigour and seriousness of the conveyance that occupies us *now* – now, not yesterday. In both cases it's a matter of evidence, of reality, yes, of objectivity, but the first of these things is indefinitely distanced, masterable only by the paths of reference, whereas the second is within our reach, and it masters us by making us close and present. We still have to learn to let ourselves be held in its grip. (REJ: 129 italics added)

The anthropologist has to learn how to 'speak well' about religious experience, to learn what religious people may hold in its grip, the beings that excite them. And she will have to convey her findings "even to those for whom 'religious matters' have become incomprehensible. And to those who believe, alas, that they understand these matters, although they have lost their interpretative key long ago" (AIME: 297).

Which religion should she choose? She set out to give alternative descriptions of the values of the Moderns, so she may be excused for limiting herself to Christianity. Her 'comparative anthropology' aims to contrast the modes of existence that define modernity, rather than compare the West with the cultures, rituals and beliefs of other peoples. But in the West, Christianity shows quite some variance. In this case, Latour's own experience will guide her – an experience he tried, "clumsy, hesitant, self-taught", to account for in *Rejoicing – Or the Torments of Religious Speech* (REJ: 174).

To study science, Latour followed Pasteur among others; to study religion, he has set up a proxy, the fictitious anthropologist, to study his own religious experience, the experience of someone who finds it difficult to convey even to his own kin what he is doing in a (Roman Catholic) church on Sundays and what matters to him (Latour 2005a: 27).

Love is a key motif in Christianity and in the language of love
– secular love, the love of one person for another – Latour suggests
one can find "a sort of prefiguring, a scale model of [the] felicity
conditions" (REJ: 118) of religious speech. If a woman asks 'do you
love me?' and her partner answers (without irony) 'yes, but you
already know that, I told you so last year',

> you'd be hard pressed to find more decisive evidence that he has
> stopped loving in earnest. He has taken the request for love as a
> request for information, as though he'd decided to carve out a path
> through space-time [. . .] to return to the distant territory of the day
> he officially declared his love. From the quality of his answer, any
> impartial observer would understand that the lover hasn't under-
> stood a thing. The fact is his girlfriend didn't ask him if he *had* loved
> her, but if he loves her *now*. That is her request, her entreaty, that is
> his challenge. (REJ: 25)

But if the lover utters the exact words he spoke last year, and once
again declares "I love you", this worn-out cliché conveys no new
information either; and

> yet she, the woman who loves, feels transported, transformed,
> slightly shaken up, changed, rearranged, or not, or the opposite,
> alienated, flattened, forgotten, mothballed, humiliated. There are
> sentences uttered every day, then, whose main object is not to map
> out references but which seek to produce something else entirely:
> the *near* and the *far*, closeness or distance. Who hasn't had some
> experience of this? (REJ: 26)

This gives a first clue to religious speech. "[It] *has no reference* – any
more than amorous exchanges do" (REJ: 28). Religious speech
does not *refer* to God, up in the sky. Religious speech doesn't refer
at all; what it does, we may learn from going back to the lovers'
dialogue.

> Let's suppose that the man understood his lover's injunction per-
> fectly well, that he has groped around clumsily for the right words
> to restate his love and found them. How are they both, after that,
> going to go around linking these new words to their initial avowal?
> They're going to say that it's the 'same' love that makes them close
> again, after a phase of distance. But naturally, this 'same' is in no
> way of the nature of a substance preserved intact over time [. . .].
> [Instead of bringing the past to the present, their experience]
> takes off from the present and goes back to the past, changing and
> deepening the past's foundation. [. . .] This reversal of the usual

> figures of time [. . .] gives them the amazing feeling that it is finally only now, *for the first time*, that they understand what has happened to them *always*. Yes, as you know very well, 'it's always the first time' – otherwise you don't love each other anymore. (REJ: 46–48)

By uttering the worn-out words 'I love you', the lover has confirmed their love, "endowing the *person* to whom his words are addressed with the existence and unity that *person* had lacked" (AIME: 302 italics added).

> Who could feel like *someone* [i.e. a *person*] without having been addressed in this way? What wretchedness, never to have aroused anything but indifference! For we don't draw the certainty of existing and being close, of being unified and complete, from our own resources but from elsewhere: we receive it as an always unmerited gift that circulates through the narrow channel of these salutary words. Our experience as recipients of such gifts is what gives us confidence to start over, again and again. (AIME: 302–303)

What the example of lovers' speech shows is that there are "words *that bear beings capable of renewing those to whom they are addressed.* [. . .] These beings have the peculiar characteristic of bringing persons from remoteness to *proximity*, from death to life. Let us say, to use more direct language, that these words *resuscitate* those to whom they are addressed – in the etymological sense, that is, they arouse them anew, get them moving again" (AIME: 303).

Transposing his observations on love-talk to religious speech, Latour writes:

> it may be astonishing to establish [. . .] a rapid bypass between the vocabulary of interactions between lovers and that of religion, quite rightly called 'revealed': what matters, however, is giving beings their true names. Throughout the tradition, those who bear not messages but tumults of the soul have been called *angels*. [. . .] [T]he prime example [has been] painted and sculpted tens of thousands of time: that of the angel Gabriel whose address comes not only to overwhelm young Mary's soul but to make her give birth to life itself. There is no better way to define beings linked to a particular type of word capable of converting those to whom they are speaking. By greeting them, they save them and impregnate them. No one has ever been able to define the soul or even decide whether or not it exists, but no one can deny that this 'whatever-it-is' lurches and vacillates in responses to such words, to such mutual overwhelming. (AIME: 303–304)

To love is more than to care for; "to care for is not to save" (AIME: 304). Once Latour has distinguished love-talk from other forms of producing subjectivities and interiority – love-talk addresses a *person*, not a psyche – he has taken a next step, from 'person' to 'soul', to 'to save' and 'to be overwhelmed'. He has used love-talk as a model to describe religious experience.

This puts the fictitious anthropologist in a position to describe the mode of existence of religious experience, [REL]. Again, summarized in a few phrases, its felicity condition is to 'save, to bring into presence'; its trajectory is 'engendering of persons' (cf. CB: 226–252); the beings to institute are 'presence-bearers'; its hiatuses are 'breaks in time'. It's a mode of existence with a key that directs us in exactly the opposite direction than [REF]. No, religious speech does not refer to some remote entity, 'God' – or in the model: the love the lovers *had* – but it makes something, a person, love, present; it directs us not backwards, but forwards, to renewal, to the 'end of times'.

The experience of love is the empirical grounding for [REL]. But along the way, this experience has been described in words that have lost meaning for at least part of the Moderns, the secularized ones. But along the same route, Latour has also given those Moderns who still consider themselves to be religious persons an alternative description of their experience. So he sends a double message. To secularized Moderns: this is what religious persons experience, this is the mode of existence of religion; to religious Moderns: this is an alternative description of what you experience; it may provide a better way to account for it than the ones that go around.

In the vocabulary of secularized Moderns, there is no place for 'souls' and 'angels'; the verb 'to save' they associate with the fire brigade, or their hard disk, not with love. For secularized Moderns, Latour is making two claims: in the first place, if you want to understand what religious people experience, think about love-speech; in the second place, note that love-talk is very special: it doesn't bring back, it doesn't *refer* to some substance, but love-speech *renews*, its direction is exactly opposite to that of reference. Moreover, don't think that if someone says he 'believes in God' he is referring to some mysterious remote entity for which he lacks the appropriate scientific tools to get access to. 'To believe in God' is not – say – to believe in some form of extra-terrestrial life. Secular Moderns are urged to stop *comparing* religion with science; science and religion are entirely different modes of existence that by invoking Latour's comparative anthropology can – and should – be *contrasted*. Latour does not invite secular Moderns to convert, but to make room for experiences that religious people have and that they themselves know in secular form as love.

But Latour has also an important message for religious people. They had better change their accounts of their religious experience. Religious speech does not refer back; religion has no substance that ensures its continuity, there is no God to refer to. Religion is not a 'belief' in something. "The word 'God' cannot designate a substance; it designates, rather, the renewal of a subsistence that is constantly at risk [. . .]" (AIME: 309). "This is [religion's] own enlightenment: it starts over, it begins again, it goes back to the starting point time after time, it repeats itself, it improvises, it innovates: moreover, it never stops describing itself, self-reflexively, as *Word*" (AIME: 306). It's all speech, ceaseless renewal of speech by speech itself, *reprise* par excellence. Péguy already had grasped that, in content and style (cf. ch.1). "If there is one mode of existence that ought to be at home in natural language, it is the religious mode" (AIME: 306). "Words that redress must be comprehensible, that's the first condition [of felicity of religious speech]; they must be said in the language of the person they are addressed to" (REJ: 54) – a conclusion that the Roman Catholic Church was late to adapt; only in 1965, the Vatican allowed the use of vernacular languages at mass.

Latour has yet another message for his church. It has to trim down also its ambitions: [REL] does not provide a metalanguage for other modes of existence, it is a mode of existence *next to* several others:

> The error of religion, in the Western context, was probably to make the Church take in too many modes and to establish it as a meta-institution. The Roman Empire must have weighed rather heavily on its shoulders; this is still the case today, with the rather mad idea of making religion serve as a pillar for bioethics, morality, social doctrine, canon law, the education of children, the vice squad . . . (AIME: 315–316)

And most of all, religion should never have considered itself to be a *competitor* of science. "If there is one question [religion] should not have answered, it is this one: 'And what do *you* say about remote beings?'" (AIME: 319).

"The modernist tragedy is to have been mistaken at the same moment, through a ricochet, as it were, about Science *and* about Religion" (AIME: 321). Neglecting the chains of translations and all the work that is required to access remote beings, to talk about a distant star-system or about microbes undetectable without instruments, epistemology mistook science for 'Double

Click communication'. Theology has entered the same trap. "Just as epistemology could not help define objective knowledge, theology cannot be relied on to help [us] speak correctly about salvation-bearing beings. [. . .] Double Click has struck both down with a single blow, obliging theologians to escape into belief through an erroneous conception of knowledge" (AIME: 305).

"A cascade of category mistakes [. . .] have made them unspeakable, unpronounceable, the very same beings that made their fathers speak, got their ancestors excited, led them to move mountains – and commit more than one crime" (AIME: 297). On Latour's lead, the fictitious anthropologist has learned how to speak well about them.

6.4   The modern experience: fifteen modes

In the course of the enquiry, fifteen 'modes of existence' are identified. No special meaning should be attached to that figure. It's the number of modes Latour claims is necessary to "get an image of the modern experience taken as a whole with a satisfactory *resolution*" (AIME: 479–480). If readers think something is missing, they are invited to contribute to the enquiry on the website that accompanies the book.

The fifteen modes are called [REP]roduction, [MET]amorphosis, [HAB]it, [TEC]nology, [FIC]tion, [REF]erence, [POL]itics, [LAW], [REL]igion, [ATT]achment, [ORG]anization, [MOR]ality, [NET]work, [PRE]position and [DC] for Double Click. Some of the names of these modes of existence are conventionally used to denote a domain. It would have been helpful if they had been chosen differently, to avoid confusingly taking domains (comprising actor-networks that for their being *set up* require a wide range of heterogeneous elements) for modes of existence (redescriptions of the values by focusing on what is *passed* on). As we have seen in the case of the pharmacologists, in a domain several modes of existence may cross over. In most cases, in studying practices what one encounters are crossings (causing potential category mistakes). Latour denotes these crossings as – e.g. – [REF.LAW].

The characteristics of each mode are summarized in a 'Pivot Table' (AIME: 488–489). For each mode, the table specifies by what *hiatus* and *trajectory* it is distinguished; what its *conditions of felicity* and infelicity are; what beings are instituted; and, finally, to what alteration being-as-other is subjected.

The Pivot Table

NAME	HIATUS	TRAJECTORY
[REP]RODUCTION	Risks of reproduction	Prolonging existents
[MET]AMORPHOSIS	Crises, shocks	Mutations, emotions, transformations
[HAB]IT	Hesitations and adjustments	Uninterrupted courses of action
[TEC]HNOLOGY	Obstacles, detours	Zigzags of ingenuity and invention
[FIC]TION	Vacillation between material and form	Triple shifting: time, space, actant
[REF]ERENCE	Distance and dissemblances of forms	Paving with inscriptions
[POL]ITICS	Impossibility of being represented or obeyed	Circle productive of continuity
[LAW]	Dispersal of cases and actions	Linking of cases and actions via means
[REL]IGION	Break in times	Engendering of persons
[ATT]ACHMENT	Desires and lacks	Multiplication of goods and bads
[ORG]ANIZATION	Disorders	Production and following of scripts
[MOR]ALITY	Anxiety about means and ends	Exploration of the links between ends and means
[NET]WORK	Surprise of association	Following heterogeneous connections
[PRE]POSITION	Category mistakes	Detection of crossings
[DC] DOUBLE CLICK	Horror of hiatuses	Displacement without translation

The Pivot Table

FELICITY/INFELICITY CONDITIONS	BEINGS TO INSTITUTE	ALTERATION	NAME
Continue, inherit, disappear	Lines of force, lineages, societies	Explore continuities	[REP]
Make (something) pass, install, protect/alienate, destroy	Influences, divinities, psyches	Explore differences	[MET]
Pay attention/lose attention	Veil over prepositions	Obtain essences	[HAB]
Rearrange, set up, adjust/ fail, destroy, imitate	Delegations, arrangements, inventions	Fold and redistribute resistances	[TEC]
Make (something) hold up, make believe/cause to fail, lose	Dispatches, figurations, forms, works of art	Multiply worlds	[FIC]
Bring back/lose information	Constants through transformations	Reach remote entities	[REF]
Start over and extend/ suspend or reduce the Circle	Groups and figures of assemblies	Circumscribe and regroup	[POL]
Reconnect/break levels of enunciation	Safety-bearers	Ensure the continuity of actions and actors	[LAW]
Save, bring into presence/ lose, take away	Presence-bearers	Achieve the end times	[REL]
Undertake, interest/ stop transactions	Passionate interests	Multiply goods and bads	[ATT]
Master scripts/lose scripts from view	Framings, organizations, empires	Change the size or extension of frames	[ORG]
Renew calculations/ suspend scruples	The "kingdom of ends"	Calculate the impossible optimum	[MOR]
Traverse domains/lose freedom of inquiry	Networks of irreductions	Extend associations	[NET]
Give each mode its template/ crush the modes	Interpretive keys	Ensure ontological pluralism	[PRE]
Speak literally/speak through figures and tropes	Reign of indisputable Reason	Maintain the same despite the other	[DC]

The fifteen modes are arranged in five groups of three, according to the way they relate to the old Subject–Object division. [TEC], [FIC] and [REF] relate to what was formerly called 'quasi-objects'. They provide alternative descriptions of science, technology, and fiction. [POL], [LAW] and [REL] offer descriptions of politics, law and religion as modes of existence; they relate to 'quasi-subjects'.

The first three modes in the Pivot Table, [REP], [MET] and [HAB], are characterized as being indifferent to the Subject–Object divide: "they explore, in being-as-other, three specific and complementary forms of alteration" (AIME: 285). In many respects, they are the most abstract modes in the list. But they are essential elements for Latour's argument.

[REP] is introduced to account for the *continuity* of existents (i.e. 'beings-as-other') in spite of the risk of hiatuses that are present, always and everywhere. It is "the mode of existence through which any entity whatsoever crosses through the hiatus of its repetition, thus defining from stage to stage a particular trajectory, with the whole obeying particularly demanding felicity conditions: to be or no longer to be!" (AIME 91–92). To subsist, an existent has to reproduce itself (at some price, i.e. with effort, and also always involving risks). The question that obsessed Hamlet has got an important amendment. Time is added: the question is not 'to be or not to be', but 'to be or *no longer* to be'. That not only applies to what we conventionally call living beings; it applies to all ones. Even a stone or a mountain has to subsist to continue its existence.

[REP] is Latour's way to account for continuity, for the prolongation of existence. Under the modern Constitution, Nature, eternal substance, was supposed to account for the continuity of the material world. But for Latour there are no substances, only beings-as-other. Introducing [REP] as a mode of existence allows Latour to explicate the 'fallacy of misplaced concreteness' that Whitehead spotted as the root of the concept of matter that has guided science. 'Matter' is a concept that confusingly merges what exists with the system of coordinates that allows us to define it as an entity in space-time (Whitehead 1967 [1925]: 79). In the vocabulary that Latour introduces, 'the material world' is an amalgam of two different modes of existence, namely [REP] and [REF]. Like Whitehead, Latour also doesn't question that materialism has been a powerful guiding principle for scientific research in the past centuries. But he argues that to account for what science is and to distinguish the practice of scientific research from the pretence that allows scientists to have sole, unique access to reality, [REP] and [REF] need to be carefully distinguished.

Scientific materialists suggest that a distinction has to be made between what has been called 'primary' and 'secondary qualities', that is to say between the 'real, physical' qualities of external objects and the qualities they excite in consciousness (like taste, odours and colour). The distinction has troubled philosophers ever since it was introduced in the early seventeenth century – 'colour' being a main battleground. Being grafted on the subject/object (and mind/matter) division, the distinction is of course not acceptable for Latour, Whitehead and James.

How then to account for experiences that under the modern Constitution are viewed as psychogenetic, that is, to be aroused 'in the mind' – not only colour and taste, but also anxiety, alienation, possession, crisis and other feelings? For this purpose, Latour introduces another mode of existence, [MET], again defined, as all other modes, in terms of hiatus, trajectory, felicity conditions, beings to institute and alteration. Expressed in common sense terms, its hiatuses are emotional shocks and crises. After an emotional crisis, one continues one's life, but in a radically different form. One has become excited, or terrified, one has become 'a different person'. However, emotion is a term we use mainly for human beings. Latour uses [MET] in a more general sense to account for whatever mutation or transformation an existent overcomes in order to subsist. So [MET] institutes not only beings we conventionally call psychogenic (commonly viewed as originating exclusively 'internal', in one's mind), but also external influences and even divinities that make an existent do (or experience) something – beings that under the modern Constitution are conceived as 'unreal' phantasies and illusions.

Introducing [MET] is an important step in abandoning the distinction between us, Moderns, and them, the Others. The "modernist believes that the others believe in beings external to themselves, whereas he 'knows perfectly well' that these are only internal representations projected onto a world that is in itself detached from meaning" (AIME: 187). The Moderns are advised to have a more critical look at themselves. The same Modern who has ridiculed the Others for believing in external forces that move them, may subscribe to a romance magazine, stuff himself with downers, or explain after a personal crisis 'I don't know what got into me.' "[T]he infrastructure that authoritizes [Moderns] to possess a psyche seems to escape them completely" (AIME: 188). Psychology is wrong in claiming that our feelings and emotions, whatever may arouse us, originate 'from inside'. Hutchins' (1995) *Cognition in the Wild* and a case study of the practice of a French ethnopsychiatrist, Nathan (CM), point to richer and more fruitful approaches.

The third mode of existence of the first group, [HAB], lets us continue in the way a key has merely indicated, and to (mis)take existence for essence. What under the modern Constitution is conceived as 'essence', Latour accounts for as habits acquired. [HAB] is "the most common [mode of existence], the most familiar of all, the one that William James [. . .] designated with the only word that fits it perfectly: habit [. . .]" (AIME: 265). Habit allows us for example to turn the page of a book without forgetting that we are reading a novel (i.e. are dealing with a text in [FIC] mode), rather than a scholarly treatise (a text in [REF] mode).

The three modes of existence of the first group, [REP], [MET] and [HAB], "are the ones that have been at most elaborated by the other collectives [the ones that used to be called 'premodern'] and most ignored by our own" (AIME: 288). By introducing them Latour has broadened the platform for discussions between the Moderns and other peoples and for that purpose they are fundamental for the overall purpose of Latour's *Inquiry*. They allow us to radically distance ourselves from the implied metaphysics of the modern Constitution and – implicitly – also from some of Latour's followers who continued to search for something with autonomous reality, independent of modes of existence and behind experience (e.g. Harman 2009: Part 2). Everything, including what the modern Constitution conceived as substance and mind, is now conceived of beings-as-other that have to articulate themselves, to translate and mobilize other beings to subsist. [REP], [MET] and [HAB] account for the fact *that* there is continuity of existents, *that* existents may metamorphose into something else and *that* we conceive continuity habitually as a sign of essence.

Why does none of these three modes have full equivalents in the Moderns' worldview? Latour argues that the Moderns have invented specific modes that unfortunately prevent them from adequately detecting them. For example, by introducing modern science, they were lured into thinking that it gave unique access to the natural world as it really is, that is, to know or at least to approximately know essences. By introducing [REP], Latour succeeds in separating what is *specific* and *valuable* about modern science (it allows reaching remote entities, by instituting reference) from this *pretence*. What scientific statements refer to are not 'brute facts' of nature, but a crossing of [REP] and [REF]. Likewise, introducing [MET] allows discussion about what other beings may move people, without the pretence that modern psychology has found the unique way to account for e.g. emotions. Introducing the three modes of the first group of the Pivot Table thus allows a diplomatic exchange with other peoples on a fair basis, without burdening the discussions with the idea that anything that doesn't fit the Moderns' *Weltanschauung* has to be considered a priori as being backward. "In Western civilization, and in Western civilization only, cultural phenomena have appeared which (as we like to think) lie in a line of development having *universal* significance and value," Weber (1972c [1922]: 9) wrote. The function of the first three modes of Latour's Pivot Table is to provide an antidote against this pretence.

The three modes of existence of the last group in the Pivot Table are also essential for Latour's enquiry, rather than redescriptions of what the Moderns value. [NET], the mode of existence of actor-networks that allows extension of associations, and [PRE], the mode of existence that institutes interpretative keys, are not outcomes of the anthropologist's investigation, that is, redescriptions of values the Moderns hold dear, but modes that allow her to start her enquiry. [NET] allows her to go around freely, having extricated herself from the notion of 'domain'; [PRE] she observes in category mistakes. The crossing [NET.PRE] provides the 'raw material' at the start of her investigation. In contrast, [DC] stands for a value the Moderns – believing that transfer of information without translation is possible – clearly adhere to, but it indicates what should *not* be done if one sets out to inquire experience. Is [DC] a proper 'mode of existence' at all? For its hiatus, Latour notes 'Horror of hiatuses' – a sure sign of its incongruousness in this list.

The Pivot Table shows that the enquiry has gradually undergone a shift. What started out as an anthropological, empirical investigation aiming to provide alternative versions of the values the Moderns hold dear (but that they can only account for confusingly), has developed into a system that also displays modes of existence, and thus values, that the Moderns ignore or have no place for. Has Latour come under the spell of systematics, has the empirical philosopher turned into an old-fashioned system builder? Latour denies it. He claims that the fifteen modes are just what he has found in the course of his enquiry of more than twenty-five years. Future research may show there are more modes, or may correct the specifications of the modes he detected. The ordering of the modes in the Pivot Table in groups only serves "compatibility with what the Moderns think about themselves" (Latour and Marinda 2015: 77). But it surely opens up modes of existence that account for experiences for which the Moderns' vocabulary have *no* proper place. They are fundamental for Latour's enquiry, both for getting the enquiry started ([NET] and [PRE]) and to prevent it from derailing ([DC]), while [REP], [MET] and [HAB], by abandoning the implied essentialism of the Moderns' worldview, open up the space for a diplomatic conversation with the others on a broader platform than the Moderns' *Weltanschauung* suggests.

Giving the Pivot Table another look, we may note that something seems to missing. If the values of politics, law and religion are redescribed in terms of [POL], [LAW] and [REL], that is, as distinct modes of existence, shouldn't we expect a separate mode of existence to account for the economic value of goods and services, a value that circulates in almost all actor-networks and that is passed through with utmost attentiveness – a value that many Moderns seem to consider to be the highest of all and that is instituted by a being that they all know very well indeed: money? Why is [ECON]omy excluded from the Pivot Table?

Latour provides two separate kinds of reasons for not including [ECON] in his Pivot Table. In the first place, his fictitious anthropologist detects among the Moderns increasing concerns that are expressed as "the most important questions we can address in common, because they concern the whole world, all humans and all things" (AIME: 386) now that we have discovered that "the horn of plenty" (AIME: 409) is emptied out by producing and distributing goods, up to the point where we need more planets to satisfy all the humans, while we only have one, the Earth. However, the anthropologist notes that the Moderns have no protocol or assembly to discuss these concerns. Instead of turning these concerns into moral scruples, the Moderns believe that "it suffices to calculate" (AIME: 387). The fictitious anthropologist has to account for the fact that for Moderns, as Jameson (2003: 76) remarked, "it is easier to imagine the end of the world than to imagine the end of capitalism."

Latour's second reason for not identifying a separate mode of existence to account for economic value runs parallel to the reasons he set out in *Reassembling the Social* to abandon the concept of 'Society'. Like 'Society', also 'the Economy' mixes up process – a specific way of assembling collectives – and the outcome of these processes. Referring to the work of Callon, MacKenzie and others on markets (cf. ch. 4), actor-network theory gives Latour's fictitious anthropologist the language to account for the economy as process, rather than as something already given as hard economic facts determined by the laws of the market.

These two reasons put the anthropologist on a path that leads to accounting for what the Moderns call 'the (capitalist) Economy' as "a contrast drawn together by *three modes of existence* that the Moderns have blended" (AIME: 385), the three modes of the fourth group of the Pivot Table, [ORG]anization, [ATT]achment and [MOR]ality.

What are these modes? To account for the role of calculation and (economic) valuation of goods, *Reassembling the Social* has already provided the necessary groundwork, while Callon, Mac-Kenzie and others have explored the empirical details. Actor-network theory describes organizing in terms of oligoptica, scripts, plug-ins and scale. They help to overcome the disorders – the hiatuses – that organizing production, distribution and consumption requires, and allow 'acting at a distance'. What people 'make do' and get interested in is accounted for in terms of *attachments* (cf. § 4.2). Shifting focus from the setup of actor-networks to what is passed through them, in *An Inquiry into Modes of Existence*, the experiences of organizing and passionate attachment to goods are accounted for by separate modes of existence, [ORG] and [ATT], characterized each by their own hiatuses, trajectory, felicity/infelicity conditions, beings to institute and alteration (cf. the Pivot Table).

But what holds all of this together? What makes 'the Economy', the container of all of this, "the metadispatcher, the 'whole is greater that its parts' that her informants [. . .] designate as the general framework within which all actors reside?" (AIME: 444) It is the Moderns' belief that "the mechanism of the market [is] asserting itself and clamouring for its competition", the "certainty of an optimum finally calculated by a higher agency that aggregates and unifies all the scripts" (AIME: 450) – a belief that originated in the nineteenth century when

> [s]cholars proclaimed in unison that a science had been discovered which put the laws governing man's world beyond any doubt. It was at the behest of these laws that compassion was removed from the hearts, and a stoic determination to renounce human solidarity in the name of the greatest happiness of the greatest number gained the dignity of a secular religion. (Polanyi 2001 [1944]: 106–107)

This makes the anthropologist sceptical: has the discipline of economics established something that no natural science can provide, namely 'certainty'? No hesitation, whatsoever? Where is morality?

Do we really need more morality? Latour's world is already full of it. Each mode of existence has its own normativity; every instauration implies a distinction between speaking, acting, articulating well or badly. "In short, all the modes participate in what could be called the institution of morality – if there had ever been such a thing" (AIME: 453). And yet, there is "a supplementary sense of good and bad that would explain the nuance to which we all seem to hold under the rubric of 'moral experience'" (AIME: 454). As being-of-other,

> an enigma is posed to every existent: 'If I exist only *through the other*, which of us then is the *end* and which is the *means*? I, who have *to pass by way of it*, am I its means or is it mine? Am I the end or is it my end?' [. . .] That tree, this fish, those woods, this place, that insect, this gene, that rare earth – are they my ends or must I again become an end for them? (AIME: 454–455)

Another hesitation, another hiatus has to be overcome: the anxiety about means and ends. So, Latour introduces another mode of existence [MOR], one that explores the links between ends and means (its trajectory), that has its own conditions of felicity (rethink, renew the calculations) and infelicity (suspend scruples), and that institutes a being that Latour designates with the term Kant used in one of his formulations of the categorical imperative – 'the kingdom of ends' –, but that is no longer limited to human beings. Finally, the specific alteration of [MOR] is formulated: "everything must be combined insofar as possible even though everything is incommensurable" (AIME: 461). Moral beings urge us to reframe, to recalculate, to start a discussion about means and ends.

There is no 'Economy', there is no economic realm any more than there is Society or Nature. There is no invisible hand that does the calculations for us to determine what is optimal, no 'metadispatcher' that distributes and allocates action. Polanyi was right to call the Moderns' belief in 'the Economy' a 'secular religion'. "There is no metadispatcher, it's as simple as that. That God, at least, does not exist; no one has ever been able to occupy that position, whether it is called Market or State. No one has ever had that kind of knowledge, that prescience. There is no Providence" (AIME: 470). To account for the experiences that started off the enquiry into economy, all three modes of existence belonging to the fourth group of the Pivot Table are brought to the fore. The value that at first look is central to the Moderns is deconstructed as an amalgam of three modes of existence, to allow an understanding of "the unpardonable crime of Capitalism" (AIME: 384).

6.5   Facing 'Gaia'

Having followed Latour's argument patiently for almost 500 pages, the pragmatic reader may wonder whether the redescriptions of values that *An Inquiry into Modes of Existence* offers will indeed help us to get a clearer sense of who we are and of the world we live in. Will the Moderns accept and adopt the redescriptions of their values that Latour offers? Not only the complexity of the *Inquiry*'s analysis should temper expectations. There surely are issues that wait for further investigation.

In the first place, we may note that the evidence on which the analyses are based varies. For his redescriptions of science ([REF]) and [TEC]nology, Latour can point to his own empirical studies and those of dozens of other scholars; for [ORG] to the work of Callon, Law and MacKenzie; [LAW] is based on an ethnography of one specific institution; [MET] among others on a study of an ethnopsychiatric practice; [REL] on Latour's articulation of his own religious experiences. In other cases, the analyses rely on everyday experiences and speculative arguments. Further detailed empirical studies of for example political practices will likely lead to amendments. And is Latour's redescription of economic values in terms of [ORG], [ATT], and [MOR] convincingly enough to turn the Moderns into agnostics where the Economy is concerned (AIME: 470)? More detailed studies of economic practice and reasoning would surely be welcome.

What about other values? For example, does what Latour writes about [REL], love and personhood cover the Moderns' values of individualism and identity (cf. e.g. Taylor 1989) – or should these values be reconstructed in terms of crossings of [REL], [MET] plus the detrimental effects of [DC]? How to deal with the modern, liberal value of avoiding cruelty and pain (Rorty 1989) – is it adequately covered by scruples about means and ends, that is, [MOR]? *An Inquiry into Modes of Existence* is explicitly presented as a preliminary report. That there are open questions that need further investigation is to be expected.

A more serious barrier to the Moderns' acceptance of the redescriptions of their values is the modern Constitution that it seeks to replace itself. Having evolved for centuries, it is deep-seated in modern consciousness and securely anchored in official institutional divisions. When Latour's analyses of science started to appear, many scientists were shocked to the point of accusing Latour of presenting a discreditable version of their values and vocation. The overwhelming force of the modern Constitution also shows in the fact that even scholars who cited Latour favourably often subsequently turned to social constructivist accounts of the 'material aspects' of social practices – an intellectually less demanding project that leaves the "testimony of the nonhumans" untouched. Meanwhile, politically engaged scholars were disappointed because – by offering redescriptions rather than social explanations – Latour's social theory did not help them with identifying the leverage-points in society for advancing their political ends.

So why one would one strain one's bookshelves with yet another heavy volume? Haven't we read already enough philosophical and sociological analyses of who we are, what has happened to us, of the limits of instrumental reasoning and the misdeeds conducted under the name of Capitalism?

All political and intellectual movements emerge from dissatisfaction with some situation. But salvation from evil is possible, their intellectual leaders promise. To change the order of being, we only have to recognize some deeper truths. Once these anchor points have been identified, change, salvation, lies in the realm of human action. Voegelin (1968) identified this inclination as a 'gnostic' attitude, as an *ersatz* religion, and illustrated its reasoning – among other things – with analyses of More's *Utopia*, Hobbes' *Leviathan* and Hegel's philosophy of history. He observed that in all cases an essential element of reality was ignored. The inhabitants of More's *Utopia* are supposed to have left man's lust for possessions behind them; Hobbes failed to observe that people may be driven by other passions than power; Hegel forgot that history wends its way into the future without our knowing its end. Every political and intellectual movement has a blind spot.

The blind spot of almost all contemporary critical theory is that it leaves the content and procedures of modern science and technology unanalysed, to only discuss the social impact of technology and to deplore the scientification of politics and the limitations of instrumental reasoning. Latour's philosophy stands out precisely on this point. It offers a way to analyse the content and role of science and technology and the role of the "testimony of nonhumans" by way of empirical, anthropological and historical, investigations – not to relativize the value and achievements of science and technology, but to specify their particular form of veridicity. Subsequently, in *An Inquiry into Modes of Existence* the same empirical-philosophical approach was introduced to analyse other modern institutions in terms of modes of existence. Each mode of existence is conceived as valuable in its own terms, each requires its own form of attentiveness, each is shown its own fragility, its own hiatuses that have to be overcome to truly manifest its worth, and its beings to institute. This allows *contrasting* values as different modes of existence, rather than *comparing* them and conceiving them as jealous competitors of science.

Most contemporary critical theories have another blind spot. Articulated within the confines of the modern Constitution, they address what are conceived of as problems of modern Society. Beyond question, these problems present themselves in stunning abundance and they require empirical research, philosophical reflection and political action. But the times are changing. Apart from abundant 'social' problems, the world is facing ecological problems of incomparable scope. Critical theory has little to say about these problems. If discussed at all, they are dressed up as challenges for society and politics. 'Nature' is supposed to lie waiting, inactively, for whatever purpose humans may decide to use it, either as a reserve for scientific research and economic exploitation, or as a dumping ground that urgently needs to be cleaned up.

On this point, too, Latour's work stands out. He too offers contributions to current debates about political issues in newspapers and weeklies (e.g. Latour 2015b; 2015c). But he thinks the real issue we have to address is another one. For instance, commenting on the horrors of the terrorists' attacks in Paris, 13 November 2015, he wrote:

Armed fanatics are criminals, no question, but they hardly jeopardize the way we live, think, produce, learn or inhabit space. We need only defend ourselves against them. But nothing in their ideology jeopardizes our deepest-held values, no more than pirates threaten the values of international trade. We have to fight against them, and that's all there is to it. [. . .] This situation has nothing to do with the civil wars of times past that divided from within. This kind of thuggery is a law-and-order matter, not war, despite all the flag-waving and calls to arms.

It's a very different story when it comes to climate change. Global warming threatens all states in every way: from industrial production, business and housing to culture and the arts. It threatens our values at the deepest level. Here is where states are actually at war with each other, battling for market share and economic development, not to mention the soft power of culture. And each of us feels divided against ourselves. If indeed there exists a 'clash of civilizations,' then this is it, and it concerns each and every one of us. (Latour 2015d)

Ever since *We Have Never Been Modern*, ecological problems have been a major concern for Latour. In *Politics of Nature* he reformulated 'political ecology' and the role of science in democracy. In *An Inquiry into Modes of Existence*, he raised the stakes.

One didn't have to be a genius, twenty years ago, to feel that modernization was going to end, since it was becoming harder and harder by the day – indeed, by the minute – to distinguish facts from values because of the increased intermixing of humans and nonhumans. At the time, I offered a number of examples, referring to the multiplication of 'hybrids' between science and society. For more than twenty years, scientific and technological controversies have proliferated in number and scope, eventually reaching the climate itself. Since geologists are beginning to use the term 'Anthropocene' to designate the era of Earth's history that follows the Holocene, this will be a convenient term to use from here on to sum up the meaning of an era that extends from the scientific and industrial revolutions to the present day. If geologists themselves, rather stolid and serious types, see humanity as a force of the same amplitude as volcanoes or even of plate tectonics, one thing is now certain: we have no hope whatsoever – no more hope in the future than we had in the past – of seeing a definitive distinction between Science and Politics. (AIME: 9)

The introduction of the notion of the Anthropocene in scientific circles marks a shift in thinking about the relationship between humans and the Earth – a shift Latour has been arguing for for decades. The catastrophic effects of climate change the IPCC projects give urgency to further exploring this shift. In the light of "planetary negotiation that is already under way" we need to reflect on all the "values that the notion of modernization had at once revealed and compromised" (AIME: 17).

Unquestionably, we have to explicate these values to other peoples. But what about being summoned to appear – as Latour declares (AIME: 9) – also before 'Gaia'? Who or what is this strange figure?

The name 'Gaia' comes from Greek mythology. The novelist William Golding suggested it to the British scientist James Lovelock as a catchy name for his hypothesis that living organisms regulate the terrestrial atmosphere – more specifically: the balance between the levels of oxygen and carbon dioxide that makes life on Earth possible. Having been criticized for conceiving the biosphere, rather than organisms, as the unit of natural selection, Lovelock later corrected his hypothesis to develop the Gaia theory. "Briefly, it states that organisms and their material environment evolve as a single coupled system, from which emerges the sustained self-regulation of climate and chemistry at a habitable state for whatever is the current biota" (Lovelock 2003: 769). Nevertheless, Lovelock continued to speak in metaphorical terms about Gaia, 'the living Earth', adding that "[l]ike life, Gaia is an emergent phenomenon, comprehensible intuitively, but difficult or impossible to analyse by reduction" (Lovelock 2003: 769). If human activities – like transport, industry and large-scale livestock breeding – continue to emit too much carbon dioxide, the subtle balance between oxygen and carbon dioxide will be disturbed. Gaia, 'the living Earth', will be endangered, with apocalyptic effects.

Is this what Latour suggests we have been summoned to appear before to explicate and defend our values – something alternatively called an emergent phenomenon, a self-regulating system, a super-organism, 'the living Earth', 'Gaia' – a mixture of British science and Greek mythology? And how should we appear before her? By performing dances that esoteric New Age movements have invented to celebrate Gaia, perhaps?

If his theory has a point, Lovelock lacks the language to express it clearly. So, in *Face à Gaïa* (2015), Latour set out to deliver a more favourable reading of Lovelock's argument, one that neither (in contrast to the concept of a 'self-regulating system') *under*-animates its central idea, nor (in contrast to the idea that the Earth is some 'super-organism') *over*-animates it (FG: 117).

To do so, he first points to a remarkable similarity between Pasteur's and Lovelock's reasoning (FG: 118 ff.). Defying Liebig and other chemists, Pasteur showed that chemistry alone could not account for lactic fermentation; some unknown *agent* was active too, which Pasteur subsequently identified as 'lactic yeast', a microbe (cf. § 2.3). Similarly, Lovelock defies the idea that the relatively stable balance between the levels of oxygen and carbon dioxide in the terrestrial atmosphere can be explained by chemical reactions alone. So, Lovelock argues, to maintain this balance something other than chemical reactions must play a role. Both Pasteur and Lovelock argue that the world is furnished with more than what the chemists thought; and to explain the phenomena they are interested in, both of them point to some previously unknown *active* agent.

What is the active agent that Lovelock points to? What keeps the balance between oxygen and carbon dioxide that makes life on Earth possible? Lovelock identifies it as the system of planetary life that comprises everything that influences the biota and is influenced by it, a self-regulating system, a 'super-organism' he called 'Gaia'. But as Latour points out in the second step of his re-interpretation of Lovelock's argument (FG: 128 ff.), in doing so, Lovelock mixes up two levels. Like the sociologists of the social who mix up two different meanings of 'the social' and who take 'Society' for a macro-entity that coordinates and constrains social action, and like economists who speak about 'the Economy' as a macro-coordinator for economic transactions, by speaking of a 'super-organism', Lovelock, too, has introduced a 'metadispatcher', a macro-entity that is supposed to have the task of assembling, coordinating and constraining the actions of a wide range of entities.

Having introduced actor-network theory to eliminate the illusion of metadispatchers from sociology and economics, we cannot expect Latour to accept another one, introduced in perhaps even more ambiguous terms. As a metadispatcher, Gaia has to go. What this badly chosen name captures is a complex assembly of many entities that each in their own interests actively adapt themselves to their environment and that try to change – translate – other entities to overcome hiatuses and to continue to exist. That jumble is what Lovelock has called 'Gaia'. There is no 'super-organism', no metadispatcher, no underlying design. There is a collective made up of heterogeneous entities that translate each other and that are translated by others, a collective that we conventionally call the Earth, a collective that includes humans.

By speaking alternatingly about a 'self-regulating system' and a 'super-organism', Lovelock has both under-animated and over-animated what according to Latour is at stake. Like Pasteur he has pointed out that to account for a phenomenon (lactic fermentation in Pasteur's case; the relatively stable balance between oxygen and carbon dioxide in Lovelock's one) we have to accept that *active agents* are part of the world. But when Lovelock falls back on modernity's scientific worldview by conceiving the collective of these entities as a self-regulating system that may be disturbed by human activities, Latour makes a full stop; there is no 'metadispatcher'. What there is, is a plural collective furnished with beings-as-other – "beings being capable of worrying us", beings that incite the question central to [MOR] about means and ends, the question "how everything can and must be combined, even though everything is incommensurable" (cf. AIME: 461). We should not have the illusion that an appeal to 'Gaia' can unify us, or represent some overarching common interest of humans and nonhumans. There simply is no overarching authority, no metadispatcher, no Providence. There is only one Earth, but Gaia is not One (FG: 130). The Earth is a plural collective. The question we have to face is how this *plural* collective can be turned into a *common* one.

Here lies the real importance and radicalness of Latour's philosophy. Latour has moved pluralism from the level of cultures to ontology, to the plurality not of how we may *see* the world, but to the world itself. With William James, Latour conceives the world as a 'pluriverse', not a universe. That move has not only radical philosophical consequences (e.g. abandonment of the idea that behind variety of appearances, there are essences), but also political (in a broad sense) ones.

Plurality invites conflict; to avoid conflict, we tend to appeal to some commonly acknowledged authority. That authority may be a God or divinities; Hobbes' sovereign; Kant's Reason; Adam Smith's invisible hand; or Nature, which ecologists appeal to. In Latour's plural world, such an appeal to an all-encompassing authority is no longer possible. Not in theory, but also no longer in practice. Appeals to God divide, rather than unify us; the sovereignty that Hobbes invented to avoid religious wars is limited to national borders and a real barrier to addressing the global ecological problems the world is facing now; for good reasons, Reason has been historically and locally relativized; there is no hidden hand that will do economic calculations for us to save us from our scruples; and also Nature does not speak with one voice – its spokespersons, scientists, disagree and are uncertain. So we would be ill-advised to look further for one overarching authority. We have to become aware of the radically conflicting nature of the problem that climate change presents to us. We should not look up to expect some overarching authority that will unify us; we need to look down, to the one Earth we inhabit with many other beings.

At the advent of modern science, pointing his telescope to the moons of Jupiter, Galileo led us "from the closed world to the infinite universe" (Koyré 1968 [1957]). Latour redirects our view back to Earth (FG: 113 ff.). Generals may argue that to win a war one needs 'boots on the ground'; Latour argues that philosophically and politically we need to get our feet on the ground, on soil, on Earth; we should not reach out for the starry heavens, but focus our attention on the planet we inhabit.

IPCC scientists have pointed to the potential catastrophic effects of climate change. They will not only wreck societies and endanger human life; it puts the Earth – that enormous collective humans are part of and live by – at stake. To discuss the risks of climate change a 'politics of the Earth' (what in *Politics of Nature* is called 'cosmopolitics') will be required. It demands a broader definition of politics than is implied in conferences of representatives from nations, convened by the UN to negotiate the measures to reduce CO_2-emissions. Nonhumans have to be represented as well, and the relations between humans and nonhumans will have to be represented in other ways than only by delegates who speak for the interests of humans divided along national borders. In *Politics of Nature*, Latour has outlined the principles of the politics that he thinks will be needed.

To put the argument to a test, anticipating the Paris Climate Change Conference COP21, Latour organized a simulation of the kind of assembly that he thinks is required. In May 2015, 200 students gathered in Nanterre, to approximate the cosmopolitical process. Apart from those who represented nations, other students represented non-national collectives (for example 'indigenous peoples') and nonhuman ones (for example 'oceans' and 'soil'), to explore alternative ways to unblock the climate negotiations. That, in practical terms, is what "being summoned to appear before Gaia" and to face her means.

Looking back on the twentieth century, Hobsbawm (1994: 287–288) wrote that the prefixes that are widely used to characterize our current – postindustrial, postmodern, post-whatever – condition, "[l]ike funerals, [take] official recognition of death, without implying any consensus or indeed certainty about the nature of life after death." Latour has not declared modernity dead, and he does not speculate about its future, nor does he offer a utopia that lies ahead. He argues that we badly need another image of ourselves and of the world. Only then, may we make public what we have been doing so far "under the table" and only then can we seriously engage in the kind of politics that is needed to turn the plural world into a common one.

Uncertainty reigns; what will happen to us, and to what extent we will succeed in achieving a common, yet plural world, depends on how we will proceed. *An Inquiry into Modes of Existence* suggests a platform for the diplomatic exchanges that will be necessary, if war, violence and ecological catastrophe are to be avoided. Its purpose is to offer the Moderns a clearer view of themselves, one that will allow them to present themselves to other peoples and, by engaging in cosmopolitics, to confront the others, with self-confidence and civility, without arrogance, to answer the question posed in 1930 to Gandhi: this is our civilization, what do you think of it? And what about yours?

Gandhi's remark should remind us that presenting our civilization, by giving "a positive, rather than a mere negative, version of those who have never been modern" (AIME: xxvi), is a delicate task. Weber argued that rationalization led to a disenchanted worldview. Latour disenchants our pretences, our claims to have found – in Christianity, science, Reason, Adam Smith's invisible hand, or in universal law – an authority that for once and for all may unify the world. Instead, we need a clearer idea about who we are and how we relate to the nonhumans, to negotiate with others, to be attentive to what they say and do and relate to the one planet we share with them. Latour's empirical philosophy, the comparative anthropology he defends in *An Inquiry into Modes of Existence*, serves that purpose.

Once we have a clearer image of our values, we may also consider re-instituting them, to give our values new – less violent – practices. How? Latour does not provide a road map. So, does Latour's philosophy, like Wittgenstein's (1969 [1952]: PU §24), "leave everything as it is", to only interpret the world in various ways, whereas – as Marx's eleventh thesis on Feuerbach states – the point is to change it? No. Latour doesn't offer an extra layer that leaves everything as it is. Redescriptions are translations too, that is, ontological moves, passes *in* the world. They re-order; they establish or strengthen links, or weaken them to untie knots. Redescriptions do not just provide interpretations or knowledge *about* the world, they are moves *in* the world; they change us, as well as the world that is addressed.

In his essay on the influence of Darwinism on philosophy, Dewey wrote:

> Old ideas give way slowly; for they are more than abstract logical forms and categories. They are habits, predispositions, deeply engrained attitudes of aversion and preference. Moreover, the conviction persists – though history shows it to be a hallucination – that all the questions that the human mind has asked are questions that can be answered in terms of the alternatives that the questions themselves present. But in fact intellectual progress usually occurs through sheer abandonment of questions together with both the alternatives they assume – an abandonment that results from their decreasing vitality and a change of urgent interest. We do not solve them: we get over them. Old questions are solved by disappearing, evaporating, while new questions corresponding to the changed attitude of endeavour and preference take their place. (Dewey 1997 [1909]: 19)

Latour's philosophy clearly offers a "changed attitude of endeavour and preference", an alternative to the intellectual framework that evolved from Plato, through Descartes, to Kant – a mode of thinking that has become deeply embedded in the common sense of the Moderns, in their views on nature, science, politics, religion, and society. For sure, Latour is not the only twentieth-century philosopher who has realized that an alternative is required for what Rorty (1989: 76, 96 ff) in one broad stroke refers to as "the Plato-Kant canon". In fact, many of towering figures of twentieth-century philosophy – e.g. James, Dewey, Whitehead, Heidegger, the later Wittgenstein, Deleuze, and to some extent even Popper – have attempted to articulate alternatives to the tradition of Western philosophy and metaphysics. What makes Latour stand out from their attempts is that he has formulated not only the philosophical groundwork of an alternative *Worldview*, but also has offered the means to further *empirically* explore the world.

The urgency of the ecological problems the world is facing is undeniable. New thinking is required – slowly, not given over to the illusions of Double Click. We need a richer common sense than the modern Constitution offers. We have to leave essentialism and the obsession with epistemology that left us failing to see what we see behind. Once we have a clearer idea of our values, we may move further – hesitantly, attentively – to explore our plural world, to innovate, to do work, to experiment with re-instituting the values we hold dear. "What risk do we run [. . .]? The world is young, the sciences are recent, history has barely begun, and as for ecology, it is barely in its infancy: Why should we have finished exploring the institutions of public life?" (PN: 227–228).

References

Amsterdamska, O. (1990) 'Surely you are Joking Monsieur Latour!', *Science, Technology and Human Values*, 15 (4): 495–504.

Austin, J. L. (1976 [1962]) *How To Do Things With Words*. Oxford: Oxford University Press.

Barnes, B. and D. Edge (eds.) (1982) *Science in Context*. Milton Keynes: The Open University Press.

Barthes, R. (1988) *Elements of Semiology*. New York: The Noonday Press.

Beck, U. (1986) *Risikogesellschaft*. Frankfurt: Suhrkamp.

___ (1993) *Die Erfindung des Politischen*. Frankfurt: Suhrkamp Verlag.

Beck, U., A. Giddens and S. Lash (1994) *Reflexive Modernization – Politics, Tradition and Aesthetics in the Modern Social Order*. Cambridge: Polity.

Berger, P. L. and Th. Luckmann (1967) *The Social Construction of Reality*. New York: Doubleday Anchor Books.

Berlin, I. (1969) *Four Essays on Liberty*. Oxford: Oxford University Press.

Blok, A. and T. E. Jensen (2011) *Bruno Latour – Hybrid Thoughts in a Hybrid World*. London: Routledge.

Bloor, D. (1976) *Knowledge and Social Imagery*. London: Routledge and Kegan Paul.

___ (1999a) 'Anti-Latour', *Studies in the History and Philosophy of Science*, 30 (1): 81–112.

___ (1999b) 'Reply to Bruno Latour', *Studies in the History and Philosophy of Science*, 30 (1): 131–136.

Brandom, R. B. (2000) *Articulating Reasons – An Introduction to Inferentialism*. Cambridge (Mass.): Harvard University Press.

Çalişkan, K. and M. Callon (2009) 'Economization, Part 1: Shifting Attention from the Economy towards Processes of Economization', *Economy and Society*, 38 (3): 369–398.

___ (2010) 'Economization, Part 2: A Research Programme for the Study of Markets', *Economy and Society*, 39 (1): 1–32.

Callon, M. (1986) 'Some Elements of a Sociology of Translation: Domestication of the Scallops and the Fishermen of St. Brieuc Bay', in J. Law (ed.), *Power, Action and Belief: A New Sociology of Knowledge?* London: Routledge and Kegan Paul, 96–233.

___ (ed.) (1998a) *The Laws of the Markets*. Oxford: Blackwell.

___ (1998b) 'Introduction: The Embeddedness of Economic Markets Economics', in M. Callon (ed.) *The Laws of the Markets*. Oxford: Blackwell, 1–57.

Callon, M. and B. Latour (1981) 'Unscrewing the Big Leviathan: How Actors Macro-Structure Reality and How Sociologists Help Them To Do So', in K. Knorr-Cetina and A. V. Cicourel (eds.), *Advances in Social Theory and Methodology – Towards an Integration of Micro- and Macro-Sociologies*. London: Routledge and Kegan Paul, 277–303.

___ (1992) 'Don't Throw the Baby Out with the Bath School!', in A. Pickering (ed.), *Science as Practice and Culture*, Chicago: Chicago University Press, 343–368.

Castells, M. (1996) *The Rise of the Network Society*. Oxford: Blackwell Publishers Ltd.

Cauli, B. et. al. (1997) 'Molecular and Physiological Diversity of Cortical Nonpyramidal Cells', *The Journal of Neuroscience*, 17 (10): 3894–3906.

Collins, H. M. (1985) *Changing Order*. London: SAGE Publications.

___ (1994) 'Book review, Bruno Latour, *We Have Never Been Modern*', *Isis*, 85: 672–674.

Collins, H. M. and S. Yearley (1992) 'Epistemological Chicken', in A. Pickering (ed.), *Science as Practice and Culture*. Chicago: Chicago University Press, 301–326.

Conant, J. B. and J. Nash (ed.) (1964) *Harvard Case Histories in Experimental Science: Volume 2*. Cambridge (Mass.): Harvard University Press.

Dahrendorf, R. (1968) *Pfade aus Utopia*. München: R. Piper and Co. Verlag.

204 References

De Vries, G. (1995) 'Should We Send Collins and Latour to Dayton, Ohio?', *EASST Review*, 14 (4): 3–10.

Dewey, J. (1997 [1909]) 'The Influence of Darwinism on Philosophy', reprinted in J. Dewey, *The Influence of Darwin on Philosophy and Other Essays*. New York: Prometheus Books, 1–19.

Durkheim, E. (1968 [1901]) *Les Règles de la Méthode Sociologique*. Paris: Presses Universitaires de France.

___ (1970 [1897]) *Suicide – A Study in Sociology*. London: Routledge and Kegan Paul.

Eco, U. (1994) *Six Walks in the Fictional Woods*. Harvard: Harvard University Press.

Feyerabend, P. K. (1975) *Against Method*. London: NLB.

___ (1976) 'On the Critique of Scientific Reason', in R. S. Cohen, P. K. Feyerabend, M. W. Wartofsky (eds.) *Essays in Memory of Imre Lakatos*. Dordrecht: Reidel, 109–143.

Fleck, L. (1979 [1935]) *Genesis and Development of a Scientific Fact*. Chicago and London: University of Chicago Press.

Foucault, M. (1979) *Discipline and Punish – The Birth of the Prison*. Harmondsworth: Penguin Books.

___ (1994) *Dits et Écrits 1954–1988*. 4 vols. Paris: Édition Gallimard.

Garfinkel, H. (1984 [1967]) *Studies in Ethnomethodology*. Cambridge: Polity.

Geertz, C. (1983) *Local Knowledge*. New York: Basic Books.

Giddens, A. (1984) *The Constitution of Society*. Cambridge: Polity.

Goodman, N. (1973) *Fact, Fiction and Forecast*. New York: Bobbs-Merrill.

Greimas, A. J. and J. Courtés (1979) *Sémiotique: dictionnaire raisonné de la théorie du langage: tome 1*. Paris: Classiques Hachette.

Gross, P. R. and N. Levitt (1994) *Higher Superstition: The Academic Left and Its Quarrels with Science*. Baltimore: The Johns Hopkins University Press.

Habermas, J. (1969) *Technik und Wissenschaft als >Ideologie<*. Frankfurt: Suhrkamp Verlag.

___ (1981) *Theorie des kommunikativen Handelns*. Frankfurt: Suhrkamp Verlag.

Hacking, I. (1983) *Representing and Intervening – Introductory Topics in the Philosophy of Natural Science*. Cambridge: Cambridge University Press.

___ (1990) *The Taming of Change*. Cambridge: Cambridge University Press.

___ (1992) 'Book review, Latour, *The Pasteurization of France*', *Philosophy of Science*, 59 (3): 510–512.

Harman, G. (2005) 'Heidegger on Objects and Things', in B. Latour and P. Weibel (eds.), *Making Things Public – Atmospheres of Democracy*. Cambridge (Mass.): MIT Press, 268–271.

___ (2009) *Prince of Networks: Bruno Latour and Metaphysics*. Melbourne: re-press.

Hobbes, T. (1980 [1651]) *Leviathan*. MacPherson, C. B. (ed.) Harmondsworth: Penguin Books.

Hobsbawm, E. (1994) *The Age of Extremes 1914–1991*. London: Abacus.

Hollis, M. and S. Lukes (eds.) (1982) *Rationality and Relativism*. Oxford: Basil Blackwell.

Hutchins, E. (1995) *Cognition in the Wild*. Cambridge (Mass.): MIT Press.

James, W. (1950 [1890]) *The Principles of Psychology*. 2 vols. New York: Dover Publications, Inc.

___ (1996 [1912]) *Essays in Radical Empiricism*. Lincoln and London: University of Nebraska Press.

Jameson, F. (2003) 'Future City', *New Left Review*, 21 (May, June), 65–79.

Kant, I. (1956 [1787]) *Kritik der reinen Vernunft*. 2. Auflage. Hamburg: Felix Meiner Verlag (translation: *Critique of Pure Reason*, translated and edited by P. Guyer and A. W. Wood. Cambridge: Cambridge University Press, 1998).

Knorr-Cetina, K. (1985) 'Book review, Latour, *Les Microbes*', *Social Studies of Science*, 15: 577–586.

Knorr-Cetina, K. and A. V. Cicourel (eds.) (1981) *Advances in Social Theory and Methodology – Towards an Integration of Micro- and Macro-Sociologies*. London: Routledge and Kegan Paul.

Koyré, A. (1968 [1957]) *From the Closed World to the Infinite Universe*. Baltimore and London: The Johns Hopkins University Press.

Kuhn, T. S. (1970) *The Structure of Scientific Revolutions*. 2nd enlarged edn. Chicago and London: University of Chicago Press.

___ (1977) *The Essential Tension*. Chicago and London: University of Chicago Press.

Latour, B. (1977) '*Pourquoi Péguy se répète-t-il? Péguy est-il illisible?*' in *Péguy Ecrivain*, Colloque du Centenaire. Paris: Klinsieck, 78–102.

___ (1984) *Les microbes – guerre et paix* suivi de *Irréductions*. Paris: Éditions A. M. Métailié.

___ (1987) *Science in Action*. Milton Keynes: Open University Press.

___ (1988a) *The Pasteurization of France*. Cambridge (Mass.): Harvard University Press.

___ (1988b) 'The Politics of Explanation: An Alternative', in S. Woolgar (ed.), *Knowledge and Reflexivity: New Frontiers in the Sociology of Knowledge*. Thousand Oaks, CA: Sage Publications, Inc., 155–177.

___ (1990a) 'Postmodern? No, Simply Amodern! Steps towards an Anthropology of Science', *Studies in the History and Philosophy of Science*, 21 (1): 145–191.

___ (1990b) 'Drawing Things Together', in M. Lynch and S Woolgar (eds.), *Representation in Scientific Activity*. Cambridge (Mass): MIT Press, 19–68.

___ (1991) *Nous n'avons jamais été modernes*. Paris: Édition la Découverte.

___ (1992) 'Where Are the Missing Masses? The Sociology of a Few Mundane Artifacts', in W. E. Bijker and J. Law (eds.), *Shaping Technology/Building Society – Studies in Sociotechnical Change*. Cambridge (Mass.): MIT Press, 225–258.

___ (1993a) 'Pasteur on Lactic Acid Yeast: A Partial Semiotic Analysis', *Configurations*, 1 (1): 129–146.

___ (1993b) *La Clef de Berlin et autre leçons d'un amateur de sciences*. Paris: La Découverte.

___ (1993c) *We Have Never Been Modern*. Cambridge (Mass.): Harvard University Press.

___ (1996a) *Aramis or the Love of Technology*. Cambridge (Mass.): Harvard University Press.

___ (1996b) *Petite reflexion sur le culte moderne des dieux faitiches*. Paris: Collection Les empêcheurs de penser en rond.

___ (1996c) 'Do Scientific Objects Have History? – Pasteur and Whitehead in a Bath of Lactic Acid', *Common Knowledge*, 5 (1): 76–91.

___ (1999a) 'For David Bloor … and Beyond: A Reply to David Bloor's 'Anti-Latour', *Studies in the History and Philosophy of Science*, 30 (1): 113–129.

___ (1999b) 'On Recalling ANT', in J. Law and J. Hassard (eds.), *Actor Network and After*. Oxford: Blackwell Publishers, 15–25.

___ (1999c) 'On the Partial Existence of Existing and Nonexisting Objects', in L. Daston (ed.), *Biographies of Scientific Objects*. Chicago and London: University of Chicago Press, 247–269.

___ (1999d) *Pandora's Hope – Essays on the Reality of Science Studies*. Cambridge (Mass.): Harvard University Press.

___ (2002) 'Gabriel Tarde and the End of the Social', in P. Joyce (ed.), *The Social in Question. New Bearings in History and the Social Sciences*. London: Routledge: 117–132.

___ (2003) 'Is *Re*-Modernization Occurring – And If So, How to Prove It? A Commentary on Ulrich Beck', *Theory, Culture and Society*, 20: 35–48.

___ (2004) *Politics of Nature*. Cambridge (Mass.): Harvard University Press.

___ (2005a) '"Thou Shall Not Freeze-Frame" or How Not to Misunderstand the Science and Religion Debate', in J. D. Proctor (ed.) *Science, Religion, and the Human Experience*. Oxford: Oxford University Press, 27–48.

___ (2005b) *Reassembling the Social – An Introduction to Actor-Network-Theory*. Oxford: Oxford University Press.

___ (2008) 'A Textbook Case Revisited – Knowledge as a Mode of Existence', in E. J. Hackett et al. (eds.), *The Handbook of Science and Technology Studies*. 3rd edn. Cambridge (Mass.): MIT Press, 83–112.

___ (2010a) 'Coming out as a Philosopher', *Social Studies of Science*, 40 (4): 599–608.

___ (2010b) *The Making of Law*. Cambridge: Polity.

___ (2011) 'Reflections on Etienne Souriau's *Les différents modes d'existence*', in L. Bryant et al. (eds.), *The Speculative Turn: Continental Materialism and Realism*. Melbourne: re.press, 304–333.

___ (2013a) *An Inquiry into Modes of Existence – An Anthropology of the Moderns*. Cambridge (Mass.): Harvard University Press.

___ (2013b) 'Biography of an inquiry: On a book about modes of existence', *Social Studies of Science*, 43 (2): 287–301.

___ (2013c) *Rejoicing – Or the Torments of Religious Speech*. Cambridge: Polity.

___ (2015a) *Face à Gaïa*. Paris: La Découverte.

___ (2015b) 'Alles im Namen der Religion', *Die Zeit*, 12 February.

___ (2015c) 'Deux leçons d'un veil iman', *Le Monde*, 15 January.

___ (2015d) 'L'autre état d'urgence', *Reporterre*, 23 November. http://www.reporterre.net/L-autre-etat-d-urgence; translation: http://www.bruno-latour.fr/sites/default/files/downloads/REPORTERRE-11–15-GB_0.pdf

Latour, B., P. Jensen, T. Venturini, S. Grauwin and D. Boullier (2012) 'The Whole is Always Smaller Than Its Parts. A Digital Test of Gabriel Tarde's Monads', *British Journal of Sociology*, 63 (4): 591–615.

Latour, B. and E. Hermant (1998) *Paris ville invisible*. Paris: La
Découverte.

Latour, B. and T. Howles (2015) 'Charles Péguy: Time, Space, and
the Monde Moderne', *New Literary History*, 46 (1): 41–62.

Latour, B. and V. A. Lépinay (2009) *The Science of Passionate Interests*.
Chicago: Prickly Paradigm Press.

Latour, B. and C. Marinda (2015) 'À métaphysique, métaphysique
et demie. L'*Enquête sur les modes d'existence* forme-t-elle un
système?' *Les Temps Modernes*, 2015/1 (n°. 682): 72–85.

Latour, B. and A. Shabou (1974) *Les idéologies de la compétence en
milieu industriel à Abidjan*. ORSTOM, Sciences Humaines, Serie
études industrielles. Abidjan: Centre de Petit Bassam.

Latour, B. and P. Weibel (eds.) (2002) *Iconoclash – Beyond the Image
Wars in Science, Religion, and Art*. Cambridge (Mass): MIT Press.

___ (eds.) (2005) *Making Things Public – Atmosphere of Democracy*.
Cambridge (Mass): MIT Press.

Latour, B. and S. Woolgar (1979) *Laboratory Life – The Social Construction of Scientific Facts*. Beverly Hills and London: Sage
Publications.

___ (1986) *Laboratory Life – The Construction of Scientific Facts*. 2nd
edn. Beverly Hills and London: Sage Publications.

Law, J. (1987) 'Technology and Heterogeneous Engineering: The
Case of Portuguese Expansion', in W. E. Bijker, T. P. Hughes and
T. J. Pinch (eds.), *The Social Construction of Technological Systems*.
Cambridge: MIT Press, 111–134.

___ (1994) *Organizing Modernity*. Oxford: Basil Blackwell.

Lovelock, J. (2003) 'The Living Earth', *Nature*, 426 (18/25 December), 769–770.

MacKenzie, D. (2006) *An Engine, not a Camera – How Financial
Models Shape Markets*. Cambridge (Mass.): MIT Press.

___ (2009) *Material Markets – How Economic Agents are Constructed*.
Oxford: Oxford University Press.

MacKenzie, D., F. Muniesa and L. Siu (eds.) (2007) *Do Economists
Make Markets? On the Performativity of Economics*. Princeton: Princeton University Press.

Mandelbaum, M. (1971) *History, Man, and Reason*. Baltimore and
London: The Johns Hopkins University Press.

Marcuse, H. (1964) *One Dimensional Man*. London: Routledge and
Kegan Paul.

Mayr, E. (1982) *The Growth of Biological Thought – Diversity, Evolution, and Inheritance*. Cambridge (Mass.): The Belknap Press of
Harvard University Press.

McGee, K. (2014) *Bruno Latour: The Normativity of Networks*. London: Routledge.

Medawar, P. (1996 [1963]) 'Is the Scientific Paper a Fraud?', reprinted in P. Medawar, *The Strange Case of the Spotted Mice and Other Classic Essays on Science*. Oxford: Oxford University Press, 33–39.

Merton, R. K. (1973) *The Sociology of Science*. Chicago: University of Chicago Press.

Mills, C. W. (1999 [1967]) *The Sociological Imagination*. Oxford: Oxford University Press.

Nehamas, A. (1985) *Nietzsche – Life as Literature*. Cambridge (Mass): Harvard University Press.

Nietzsche, F. (1966) *'Aus dem Nachlass der achtziger Jahre'*, in F. Nietzsche, *Werke in drei Bände*. München: Hanser Verlag, III: 417–925.

Ornstein, M. (1963 [1913]) *The Rôle of the Scientific Societies in the Seventeenth Century*. Hamden and London: Archon Books.

Pasteur, L. (1858) 'Mémoire sur la fermentation appelée lactique', *Annales de chimie et de physique*, 3ième Série, Tome LII, 404–418 (partially translated, with comments, in Conant and Nash (eds.) 1964).

Plato (1961) 'Republic' in *The Collected Dialogues of Plato*. Hamilton, E. and H. Cairns (eds.) (1961) Princeton: Princeton University Press.

Polanyi, K. (2001 [1944]) *The Great Transformation – The Political and Economic Origins of Our Time*. Boston: Beacon Press.

Popper, K. R. (1972) *Objective Knowledge – An Evolutionary Approach*. Oxford: Oxford University Press.

Porter, T. (1995) *Trust in Numbers – The Pursuit of Objectivity in Science and Public Life*. Princeton: Princeton University Press.

Quine, W. V. O. (1951) 'Two Dogmas of Empiricism', *The Philosophical Review*, 60 (1): 20–43. reprinted (with revisions) in W. V. O. Quine (1961) *From a Logical Point of View*. 2nd edn. New York: Harper Torchbooks, 20–46.

Rorty, R. (1989) *Contingency, Irony, and Solidarity*. Cambridge: Cambridge University Press.

Ryle, G. (1970 [1949]) *The Concept of Mind*. Harmondsworth: Penguin Books.

Sartre, J. P. (1966 [1945]) *L'existentialisme est un humanisme*. Paris: Gallimard – Folio Essais.

Schaffer, S. (1991) 'The Eighteenth Brumaire of Bruno Latour', *Studies in History and Philosophy of Science*, 22 (1): 174–192.

Schmidgen, H. (2011) *Bruno Latour zur Einführung*. Hamburg: Junius Verlag.

___ (2013) 'The Materiality of Things? Bruno Latour, Charles Péguy and the History of Science', *History of the Human Sciences*, 26 (1): 3–28.

Schmitt, C. (1996 [1932]) *The Concept of the Political*. Chicago: University of Chicago Press.

Schnädelbach, H. (1983) *Philosophie in Deutschland 1831–1933*. Frankfurt: Suhrkamp Verlag.

Schopenhauer, A. (1977 [1818]) *Die Welt als Wille und Vorstellung*. Zürich: Diogenes.

Searle, J. (1995) *The Construction of Social Reality*. London: Allen Lane The Penguin Press.

Serres, M. and B. Latour (1995) *Conversations on Science, Culture, and Time*. Ann Arbor: University of Michigan Press.

Shapin, S. and S. Schaffer (1985) *Leviathan and the Air-Pump*. Princeton: Princeton University Press.

Simondon, G. (1989 [1958]) *Du mode d'existence des objects techniques*. Paris: Aubier.

Skinner, Q. (1992) 'Who are 'We'? Ambiguities of the Modern Self', *Inquiry*, 34, 133–153.

Souriau, E. (1943) *Les différents modes d'existence*. Paris: Presses Universitaires de France.

Stengers, I. (1996–1997) *Cosmopolitiques 1–7*. Paris: La Découverte.

Stenius, E. (1964) *Wittgenstein's Tractatus*. Oxford: Basil Blackwell.

Sulston, J. and G. Ferry (2002) *The Common Thread*. London: Bantam Press.

Tarde, G. (1999 [1895]) *Monadologie et sociologie*. Paris: Institut Synthélabo.

Taylor, C. (1989) *Sources of the Self – The Making of the Modern Identity*. Cambridge: Cambridge University Press.

Tolstoy, L. (2010 [1869]) *War and Peace*. Oxford: Oxford University Press.

Tournier, M. (1972 [1967]) *Vendredi ou les limbes du Pacifique*. Paris: Gallimard.

Trevor-Roper, H. (1983) 'The Highland Tradition in Scotland', in E. Hobsbawm (ed.), *The Invention of Tradition*. Cambridge: Cambridge University Press, 15–42.

Venturi, T. (2009) 'Diving in Magma: How to Explore Controversies with Actor-Network Theory', *Public Understanding of Science*, 19 (3): 258–273.

___ (2010) 'Building on Faults: How to Represent Controversies with Digital Methods', *Public Understanding of Science*, 21 (7): 796–812.

Venturi, T. et al. (2014) 'Three Maps and Three Misunderstandings: A Digital Mapping of Climate Diplomacy', *Big Data and Society*, July–December, 1–19.

Venturi, T., P. Jensen and B. Latour (2015) 'Fill in the Gap. A New Alliance for Social and Natural Sciences', *Journal of Artificial Societies and Social Simulation*, 18 (2).

Vernon, K. (1990) 'Book review, Latour, *The Pasteurization of France*', *The British Journal for the History of Science*, 23 (3): 344–346.

Voegelin, E. (1968) *Science, Politics and Gnosticism*. Washington: Regnery Publishing.

Voltaire (2006 [1764]) *Voltaire's Philosophical Dictionary*, http://www.gutenberg.org/ebooks/18569#download

Weber, M. (1968 [1919]) *Gesammelte Aufsätze zur Wissenschaftslehre.* Tübingen: J. C. B. Mohr <Paul Siebeck>.

___ (1972a [1922]) *Gesammelte Aufsätze zur Religionssoziologie.* Tübingen: J. C. B. Mohr <Paul Siebeck>.

___ (1972b [1922]) *Wirtschaft und Gesellschaft – Grundriss der verstehenden Soziologie.* Tübingen: J. C. B. Mohr <Paul Siebeck>.

___ (1972c [1922]) *Die Protestantische Ethik.* München und Hamburg: Siebenstern Taschenbuch Verlag.

Weinberg, S. (1996) 'Sokals' Hoax', *The New York Review of Books*, 8 August.

Whitehead, A. N. (1967 [1925]) *Science and the Modern World.* New York: The Free Press.

Winch, P. F. (1958) *The Idea of a Social Science.* London: Routledge.

Wittgenstein, L. (1969 [1921]) *Tractatus Logico-Philosophicus*, in L. Wittgenstein, *Schriften 1*, Frankfurt: Suhrkamp Verlag.

___ (1969 [1952]) *Philosophische Untersuchungen*, in L. Wittgenstein, *Schriften 1*, Frankfurt: Suhrkamp Verlag.

___ (1970) *Lectures and Conversations on Aesthetics, Psychology and Religious Belief.* Oxford: Basil Blackwell.

___ (1993) 'Bemerkungen über Frazers *Golden Bough*', in L. Wittgenstein, *Philosophical Occasions 1912–1951* ed. J. C. Klagge and A. Nordmann. Indianapolis and Cambridge: Hacket Publishing Company, 118–155.

___ (1998) *Vermischte Bemerkungen.* Frankfurt: Suhrkamp Verlag.

Name Index

214 Name Index

Subject Index

Page-numbers in **bold** indicate main explications of technical terms.

nature/society divide, 3, 116, 125–
128, 130, 161
network
actor-network theory concept
of -, **92**
social -, 92
nonhumans, 44–46, 48, 62, 64, 67,
70–72, 77–80, 82, 84–86, 88,
90–95, 100, 119, 121, 123–125,
127, 128–139, 142, 146, 147,
153, 173, 192–194, 197–199
have history too, 132
introduced to allow agency for
anything non-human, **40**, 88
not an odd name for nature or
material world, 88
normativity, 108, 111, 112, 171, 173,
174, 190
incorporated in ontology, 174
versus facticity, **173**
noumenal world, 30, 125

object, (main) circulating -, **41**, 43,
48, 50, 92, 106, 108
oligopticon, **96**, 97 98, 102, 189
ontology, **10**, 38, 52, 54, 63, 66, 67,
76, 80, 112, 126, 127, 129, 134,
135, 138, 156, 158, 163, 165,
166, 174, 197
another turn after the social
turn, 69, 70, 76, 78
gets pride of place over
epistemology, 130
relationist, 65, 66, 135, 136
variable ontologies, 89, 133, 135,
136
organic theory (Whitehead), 160

panorama, **97**, 144
Parliament of Things, 137, 145,
146
perplexity, requirement of -, **140**,
141–144
person, 177, 178, 180
versus psyche, 179
phenomenal world, 30, 125, 126

philosophy, 10, 11, 15, 16, 40, 65,
67, 68, 83, 108, 125, 126, 130,
135, 160, 161, 185, 192, 200; *see
also* name-index for individual
thinkers
empirical -; *see* empirical
philosophy
task of -, 3, 4
philosophy of language, 107, 109,
155, 159, 172
philosophy of law, 110–112
philosophy of science, 1, 11,
14, 22, 23, 37, 52, 68, 69,
160; *see also* empiricism,
logical -
philosophy of technology, 128
Pivot Table, 181, **182–183**, 184,
187–189, 191
plasma, 7, 11
plug-in, **97**, 98, 102, 135, 144,
189
plural, yet common world, 137–
139, 143, 145–147, 197
plurality of being *versus* – of
language, culture or views,
158, 161, 163, 197
political philosophy, 122, 136, 137,
144, 145; *see also* cosmopolitics
'green' -, 136
politics, 1–3, 101, 114, 115, 117–126,
129–131, 136, **§5.4**, 192–194,
197–200; *see also* [POL]
instituting order without
recourse to a priori
agreement, **137**
population thinking, 159
postmodernism, 21, 48, 114, 118,
165, 198
premodern, 128, 130, 186
preposition, 167, 169, 172
process of splitting and inversion,
34, 35, 49, 50
proposition (Whitehead), 134, 140,
141, 147, 158
Providence, 117, 147, 191, 197
purification, **119**, 120, 127–129